Contents

Introduction

Introduction

Welcome to this GCSE student book, which has been written specially to support you as you work through your Design and Technology course. If you are following a short course, check with your teacher to see which sections of the book you need to cover. You can find out information about the full course and the short course in the next couple of pages.

How to use this book

This book will help you develop knowledge and understanding about the specialist materials area you have chosen to study within Design and Technology. It includes four main sections on:

- classification and selection of materials and components
- preparing, processing and finishing materials
- manufacturing commercial products
- design and market influence.

As you work through each of the two-page subsections within the four main sections listed above, you will find a blue box of 'Things to do'. These tasks are designed to test your understanding of what you have learned. Your teacher may ask you to undertake these tasks in class or for homework.

At the end of each main section, you will find a number of practice examination questions. These questions are similar in style to the ones in the end-of-course exam.

In preparation for your exam, it is a good idea to put down on a single side of A4 paper all the key points about a topic. Use subheadings or bullet point lists and diagrams to help you organize what you know. If you do this regularly throughout the course, you will find it easier to revise for the exam.

The book also includes sections that cover the coursework requirements of the full course and the short course. These coursework sections will guide you through all the important designing and making stages of your coursework. They explain:

- how to organize your project
- what you have to include
- how the project is marked
- what you have to do to get the best marks.

You should refer to the coursework sections as and when you need.

The GCSE Design and Technology full course

The GCSE Design and Technology full course builds on the experience you had of all the five materials areas at Key Stage 3:

- Food Technology
- Textiles Technology
- Resistant Materials Technology
- Graphic Products
- Systems and Control Technology.

Each of the five materials areas will provide opportunities for you to demonstrate your design and technology capability. You should therefore specialize in a materials area that best suits your particular skills and attributes.

What will I study?

Throughout the full course you will have the opportunity to study:

- materials and components
- production processes
- industrial processes
- social, moral, ethical and environmental issues of product design
- product analysis
- designing and making processes.

The content of this book will provide you with all the knowledge and understanding you need to cover during the full course.

You will then apply this knowledge and understanding when designing and making a 3D product and when producing an A3 folder of design work. You should spend up to 40 hours on your coursework project, which accounts for 60 per cent of your Design and Technology course.

At the end of the full course, you will be examined on your knowledge and understanding of your chosen materials area. There will be a $1\frac{1}{2}$ hour exam, worth 40 per cent of the total marks. The exam will be made up of four questions, each worth 10 per cent of the marks.

The GCSE Design and Technology short course

The GCSE Design and Technology short course is equivalent to half a full GCSE and will probably be delivered in half the time of the full course. It involves the study of half the content of the full GCSE, and the development of HALF the amount of coursework.

The GCSE short course allows you to work in the materials area you feel best suits your own particular skills and attributes. You can choose from:

• Food Technology
• Textiles Technology
• Resistant Materials Technology
• Graphic Products
• Systems and Control Technology.

The content of this book will provide you with all the knowledge and understanding you need to cover during the short course.

You will then apply this knowledge and understanding when designing and making a 3D product and when producing an A3 folder of design work. You should spend up to 20 hours on your coursework project, which accounts for 60 per cent of your Design and Technology course.

At the end of the short course, you will be examined on your knowledge and understanding of your chosen materials area. There will be a one hour exam, worth 40 per cent of the total marks. The exam will be made up of three questions.

Managing your own learning during the course

At GCSE level, you are expected to take some responsibility for planning your own work and managing your own learning. The ability to do this is an essential skill at Advanced Subsidiary (AS) and Advanced GCE level. It is also highly valued by employers.

In order that you start to take some responsibility for planning your own work, you need to be very clear about what is expected of you during the course. This book aims to provide you with such information. Helpful hints include:

• Read through the whole of the introduction before you start the course so you fully understand the requirements of either the full course or the short course
• Investigate the coursework sections that give you a 'flavour' of what you are expected to do
• Check out how many marks are awarded for each of the assessment criteria. The more marks that are available, the more work you will need to achieve them
• discuss the coursework deadlines with your teacher so you know how much time is available for your coursework.

ICT skills

There will be opportunities during the course for you to develop your information and communications technology (ICT) capability through the use of computer aided design and computer aided manufacture (CAD/CAM). You may have the opportunity to use:

• ICT for research and communications, such as using the Internet, e-mail, video conferencing, digital cameras and scanners
• word processing, databases or spreadsheets for planning, recording, handling and analysing information
• CAD software to model, prototype, test and modify your design proposals
• CAM using computer-controlled equipment.

Understanding industrial and commercial practice

During your GCSE course, you will have the opportunity to develop an understanding of the design and manufacture of commercial products by undertaking product analysis.

You should demonstrate your understanding of industrial practices in your designing and making activities, which could include:

• developing design briefs and specifications
• using market research
• modelling and prototyping prior to manufacture
• producing a working schedule that shows how the product is manufactured
• making a high quality product that matches the design proposal
• testing and evaluating your product against the specification to provide feedback on its performance and fitness-for-purpose.

You should also use the appropriate technical words to describe your work. Many of these words are to be found in this book.

When important words or words you may not know first appear, they are in **bold**. This means that you can look up their meaning in the glossary that appears at the end of the book.

Section A:
Classification and selection of materials and components

Softwoods and hardwoods

Aims

- To understand that woods are classified as softwoods or hardwoods.
- To understand the characteristics and differences between softwoods and hardwoods.
- To understand the range and use of woods.

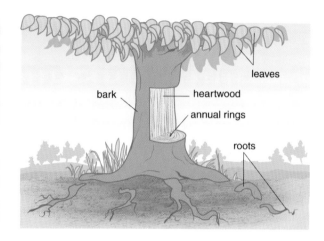

Section through a tree

The classification and structure of woods

Wood is a very versatile material. It has been used through history for many purposes: fuel for fires, weapons for hunting, cooking utensils, transportation and for structural use in houses.

There are two main types of timber (from different sorts of tree):

- **hardwoods**
- **softwoods**.

Hardwoods

The term hardwood is associated with timbers coming from deciduous (leaf losing) trees. The use of the word hardwood is not always a true description of the wood's **hardness**. Some hardwoods such as balsa wood are much lighter in weight and softer than most softwoods.

Hardwoods are generally slower growing, taking up to 100 years to reach full maturity. This makes them very expensive to buy.

Softwoods

Softwoods are commonly classified as coniferous (cone bearing). Because they are evergreen, they reach maturity in about 30 years, making them relatively cheaper than hardwoods and much more commercially available.

Leaf losing trees are hardwoods

Timber	Hard/soft	Origin	Properties/characteristics	Uses
Oak	Hardwood	Europe, USA	• Hard and tough • Durable • Finishes well • Heavy • Contains an acid which corrodes steel	• High quality furniture • Garden benches • Boat building • Veneers
Mahogany	Hardwood	Central and South America	• Easy to work • Durable • Finishes well • Prone to warping (going out of shape)	• Indoor furniture • Interior woodwork • Window frames • Veneers
Beech	Hardwood	Europe	• Hard and tough • Finishes well • Prone to warping • Turns well	• Workshop benches • Children's toys • Interior furniture
Ash	Hardwood	Europe	• Tough • Flexible (good elastic properties) • Works and finishes well	• Sports equipment • Ladders • Laminated furniture • Tool handles
Birch	Hardwood	Europe	• Hard wearing	• Plywood veneers
Pine (Scots)	Softwood	Northern Europe	• Easy to work • Knotty and prone to warping	• Constructional wood work (joists, roof trusses) • Floorboards • Children's toys

The properties and uses of timbers

Tree growth

Even though hardwoods and softwoods have different biological cell structures, they basically grow in a similar way. The amount of growth depends upon:

• the kind of tree – softwoods grow quicker
• the soil conditions
• the position of the tree – how much light it gets
• the climate – trees grow quicker in tropical countries because it is generally hot and wet.

During the growing season, normally spring through to autumn, the tree's girth (thickness) increases along with its height. In most trees, the cells produced during the drier summer months have thicker cell walls. This summer growth is mainly responsible for the **mechanical strength** of the timber.

The structure of softwoods is generally made up from tube-like cells. This normally makes softwoods less dense than hardwoods. Softwoods are also more prone to water damage. This is because the timber absorbs the water like a sponge if the end grain is exposed and left untreated.

Hardwoods contain much more **fibrous** material. The fibres are smaller and more compact which gives the wood greater mechanical strength and greater hardness.

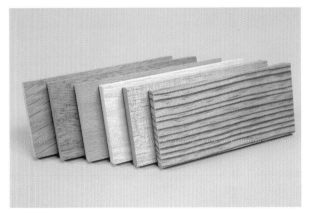

A selection of timbers

▪ Things to do ▪

1 Copy out the table and record a summary of the differences between hardwoods and softwoods. A start has been made for you.

Hardwood	Softwood
Slower growing taking 100+ years to reach full maturity	Reaches maturity after 30+ years

Manufactured boards

Aims

- To understand the range of manufactured boards available.
- To understand the advantages and disadvantages of manufactured boards.

Common shapes and available sizes of wood

Once a tree has been felled, cut down and taken to a sawmill, it is converted ready for **seasoning**. After the timber has dried out, it is cut into smaller sections of common sizes and shapes. Most timber now sold in DIY stores is **planed all round (PAR)**.

Timber for sale at a timber yard or builders merchants tends to be **rough-sawn**. When using rough-sawn timber, you must remember to allow for some waste when it is machined to make it PAR.

Manufactured boards

The amount of solid timber used today in the furniture and construction industries is small in comparison to that of manufactured boards. The **mass production** of furniture is almost entirely based around a range of manufactured boards such as:

- MDF (medium density fibreboard)
- plywood
- chipboard
- blockboard
- hardboard.

The use of such boards has both advantages and disadvantages.

Advantages

- Manufactured boards are available in large flat sheets – 2440 x 1220 mm.
- They have good **dimensional stability** – they do not warp as much as natural timbers.
- They can be decorated in a number of ways, such as with **veneers** or paint.
  - Sheets of plywood and MDF are flexible and easy to bend over formers for laminating.
  - Waste material is used to make some MDF, chipboard and hardboard.

Disadvantages

- Sharp tools are required when cutting, and tools are easily blunted.
- Thin sheets do not stay flat unless supported.
- Difficult to join in comparison to traditional construction methods.
- They must be treated to cover unsightly edges and to stop water getting in.
- The cutting and sanding of some types of board generate hazardous dust particles.

Manufactured boards are very useful in the production of furniture, but they do present problems such as joining and covering the exposed edges. The concealing of edges is covered in the diagram to the right and the joining of boards on pages 22–3 and 50–1.

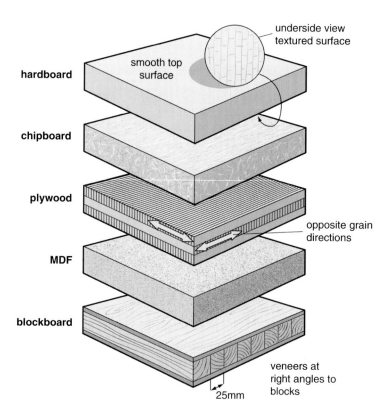

hardboard — smooth top surface — underside view textured surface

chipboard

plywood — opposite grain directions

MDF

blockboard — veneers at right angles to blocks — 25mm

Manufactured boards

Board	Characteristics	Uses	
Medium density fibreboard	MDF is very widely used in mass-produced, flat-packed furniture. It is a very dense material with an excellent surface finish, which can be veneered or painted. It is very stable and is not affected by changing humidity levels.	• Flat-pack furniture • Drawer bottoms • Kitchen units • Heat and sound insulation	
Plywood	This board is made up from an odd number of layers (veneers) normally 1.5 mm thick. The grain of each layer is at right angles to the layer either side of it. The two outside layers run in the same direction. Plywood has great tensile strength but is prone to splitting when cut. Birch veneers are usually used on the outside layers resulting in an attractive surface.	• Boat building (exterior quality plywood) • Drawer bottoms and wardrobe bottoms • Tea chests • Cheaper grades used in construction industry for hoardings and shuttering	
Chipboard	This board is made from waste products. It is bonded together using very strong resins. Although it has no grain pattern, it is equally strong in all directions but not as strong as plywood. The surface is generally veneered or covered with a plastic laminate. It is not very resistant to water although special moisture resistant grades are available.	• Large floor boards and decking for loft spaces • Shelving • Kitchen worktops • Flat-pack furniture	
Blockboard	This board has long strips of timber running down its length. A veneer is applied to the external surface.	• Fire doors • Tabletops	
Hardboard	This is the cheapest of the manufactured boards. It is made from compressed fibres that have been soaked in a resin before being compressed. One side is very smooth and the underside is textured. It is unsuitable for external use because it absorbs moisture easily.	• Drawer bottoms • Cabinet backs • Smoothing out uneven floors • Lightweight internal door cladding	

Characteristics and uses of manufactured boards

veneer

butt joint

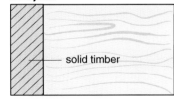

solid timber

tongue and groove

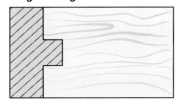

Concealing edges

■ Things to do ■

1 Why are hardwood planks available in sizes up to 300 mm wide but softwood only up to 200 mm wide?

2 Use a catalogue from a timber supplier and draw up a table of comparative costs for oak, mahogany, beech, ash, birch and Scots pine.

3 Why are hardwoods relatively more expensive than softwoods?

4 More and more mouldings, architrave and skirting boards are now being made from MDF. Why do you think this is the case?

5 Use a supplier's catalogue to identify the standard sizes of board.

6 Make a list and sketch the profile of all the pre-formed sections of timber used in a room in your home, e.g. skirting board, architrave, etc.

Ferrous and non-ferrous metals

Aims

- To understand that metals are classified in three groups: ferrous, non-ferrous and alloys.
- To understand the difference between ferrous, non-ferrous metals and alloys.
- To understand the range and use of metals.

The difference between ferrous and non-ferrous metals

Metals play a vital role in our everyday lives but we take them for granted. What would it be like without metal cutlery, pots and pans, cars, bikes, buses, trains, pipes and taps to carry our water and most buildings which now have metal frames?

Today, the study of metals – metallurgy – is a highly technical and scientific subject. Metallurgists have made metals with special properties. These metals have allowed man to travel to the moon and have enabled some people to have hip and knee replacements.

Metals are an especially ideal material for use in **mass production**. Metals can be bent, stretched, drilled, cast, riveted, welded, cut, twisted and folded – all with great accuracy. Heat treatment processes can be applied to make the metals harder or softer (see pages 34–5).

The general strength of metals is one of their major advantages. This property has enabled giant structures like bridges, ships and oil rigs to be constructed. In contrast to this, precious metals like gold and silver are crafted by skilled individuals to produce rings, goblets and products of great beauty.

Metals form about 25 per cent by weight of the earth's crust. They are found all over the earth, although they are not evenly distributed. Aluminium is the most common metal, accounting for some eight per cent, followed by iron with about five per cent.

Metals are divided into three basic categories – **ferrous**, **non-ferrous**, and **alloys**.

Modern steel chimneys

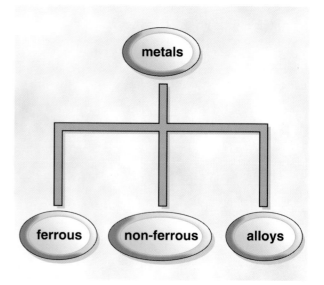

Categories of metals

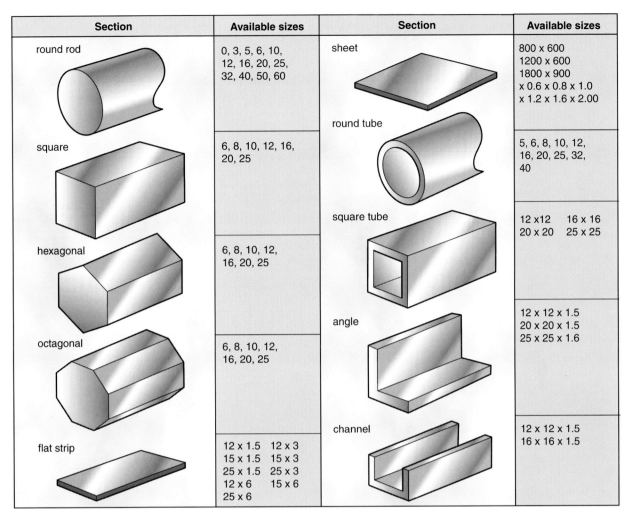

Section	Available sizes	Section	Available sizes
round rod	0, 3, 5, 6, 10, 12, 16, 20, 25, 32, 40, 50, 60	sheet	800 x 600 1200 x 600 1800 x 900 x 0.6 x 0.8 x 1.0 x 1.2 x 1.6 x 2.00
square	6, 8, 10, 12, 16, 20, 25	round tube	5, 6, 8, 10, 12, 16, 20, 25, 32, 40
hexagonal	6, 8, 10, 12, 16, 20, 25	square tube	12 x12 16 x 16 20 x 20 25 x 25
octagonal	6, 8, 10, 12, 16, 20, 25	angle	12 x 12 x 1.5 20 x 20 x 1.5 25 x 25 x 1.6
flat strip	12 x 1.5 12 x 3 15 x 1.5 15 x 3 25 x 1.5 25 x 3 12 x 6 15 x 6 25 x 6	channel	12 x 12 x 1.5 16 x 16 x 1.5

Common sections of metals

Ferrous

This group of metals is composed mainly of iron. Small additions of other elements are added such as carbon, tungsten, chromium and nickel. Almost all ferrous metals are magnetic.

Non-ferrous

This group of metals contains no iron and almost entirely consists of pure metals such as aluminium, copper, tin and lead. Non-ferrous metals are not magnetic.

Alloys

An alloy is a new metal which is formed by mixing two or more metals and sometimes other elements together. An endless list of alloys is possible all with their own individual properties. Alloys are normally grouped into ferrous and non-ferrous alloys.

Ferrous alloys	Non-ferrous alloys
Mild steel	Brass
Stainless steel	Duralumin
Cast iron	Bronze

Common shapes and available sizes of metals

As with most materials, metals are produced in a range of standard shapes and sizes, regardless of the type of metal.

The full range of sizes available can be found in a supplier's list with up-to-date prices. It is important when designing to bear in mind what shapes, sizes and profiles you have available for use, and you can use the supplier's guide as a reference at this stage. The most common sections are shown in the table above.

▪ Things to do ▪

1 Collect a range of different metals from around the workshop. State whether each one is a ferrous or non-ferrous metal.

2 Why is steel plated with tin used in the food industry?

Steel

Aim

- To understand the production of steel and its applications

The production of steel

Steel in its basic form is an alloy of iron and carbon. Iron, in the form of iron **ore**, is spread over the earth's crust. For it to be useful, the ore has to be broken down and the pure iron extracted from it. To do this the ore has to be washed, sorted out or graded, and crushed. It is then roasted and the various unwanted impurities are burnt off.

Coke and **limestone** are then added to the iron ore. Coke is used as a fuel to aid the burning process and the limestone can be used as a **flux**, helping everything to flow and mix together. Not all iron ores need limestone as a flux.

A huge furnace burns continuously and air is forced through the mixture at high pressure. This burns off the excess carbon until the precise mixture is reached. The molten metal is then tapped off and cast into slabs or large **ingots**. The ingots are then further processed or refined.

In order to make steel, the carbon content must be further reduced in the pig iron. It would normally be about 3 per cent, but for mild steel it must be below 0.35 per cent. Any remaining impurities are also removed at this stage. Other elements can also be added if required, such as 12 per cent chromium, 8 per cent nickel to make stainless steel.

Steel castings being poured

Metal	Composition	Properties/characteristics	Uses
Iron	Pure metal	• Soft and ductile • Weak in tension	• Alloyed with carbon to make steel
Mild steel	Alloy of iron and carbon (0.15 – 0.3% carbon)	• Tough, ductile and malleable • Good tensile strength • Easily joined by welding or brazing • Poor resistance to corrosion • Cannot be easily heat treated • Easily worked in school workshop	• Structural steel girders • Car body panels • Nails, screws, nuts and bolts, general ironmongery
Stainless steel	Alloy of steel (12% chromium, 8% nickel)	• Hard and tough • Excellent resistance to corrosion • Difficult to use in school workshop	• Cutlery • Kitchen sinks • Pots and pans
Silver steel	Alloy of steel (0.8 – 1.5% carbon)	• Very hard and less ductile than mild steel • Difficult to cut but easily joined by welding	• Scribers • Screwdriver blades

Iron and steel

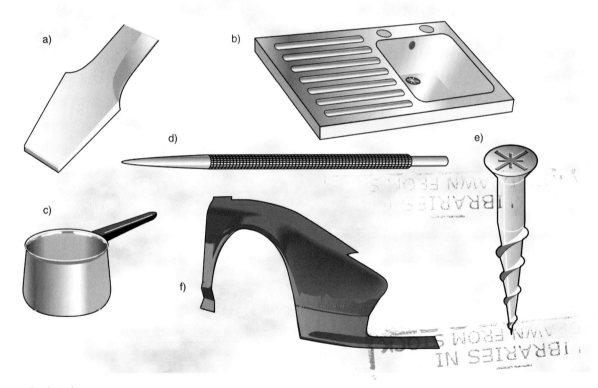

Items made of steel

Large **billets** and slabs are then rolled, water cooled and cut to length. As a result of the slabs being rolled hot, a black **oxide** finish is left on the surface. This type of steel is called **black mild steel**.

To achieve a bright surface finish with accurate dimensions, the steel is cleaned, oiled and re-rolled while cold. This steel is then drawn through a series of decreasing-sized dies (holes) to produce the accurately sized sections. This type of steel is known as bright drawn mild steel (BDMS).

■ **Things to do** ■

1 The diagrams above show items made from steel.

a screwdriver blade   d scriber
b kitchen sink   e screw
c saucepan   f car wing

Identify what type of steel each item has been made from. Give reasons for your choices.

Non-ferrous metals

Aims

- To understand the characteristics of non-ferrous metals and alloys.
- To understand the range of uses and applications to which non-ferrous metals and alloys are put.

Aluminium

Aluminium is the most abundant metal in the earth's crust. The demand for lightweight, strong metals has also made the production of aluminium the biggest in terms of output.

Aluminium is extracted from the earth in the form of bauxite. It is difficult to break down and decompose. The most common way to extract it is by using the process of **electrolysis** (it can be extracted chemically but this is even more expensive). The aluminium **ore** is ground to a powder to produce **alumina** which is then put into a bath containing a solution of chemicals at 1000°C. An electric current is then passed between the **electrodes** and the pure aluminium is deposited at the bottom of the bath (see diagram opposite). This is an expensive process and is the reason why aluminium is more expensive than steel.

Copper

Copper is mined in much the same way as aluminium. The ore is crushed and separated by **flotation**. This allows the smaller rock particles to sink and the copper grains to be carried away in the water.

The copper residue is then **smelted** in a furnace where a flux is added, helping to remove the impurities. The resulting mixture of iron sulphates and copper is further processed in a **converter** leaving the copper to be cast into large slabs.

A final refining process leaves pure copper. This process is very expensive and explains why the price of copper is more expensive than steel.

Lead and zinc

Lead and zinc are produced using similar processes, as both are derived from their sulphides. The ore is concentrated using the floatation process and then roasted to form the oxide. This is then reduced using carbon to produce the metal. As zinc has a low boiling point the zinc vapour is collected and condensed. Zinc of high purity can be produced by electrolysis

Tin

Tin ore is crushed, washed and roasted before entering the blast furnace. With the impurities burnt off, the tin is cast into large slabs. The final processing by electrolysis makes the purest form of tin.

Open cast mining of iron ore

Metal	Composition	Properties/characteristics	Uses
Aluminium (Al)	Pure metal	• Light, soft and **ductile** • Good strength for a light material • Difficult to join by soldering or welding • Good conductor of heat and electricity • Corrosion resistant and polishes well to give a good surface finish	• Window frames • Soft drinks cans • Kitchen foil • Alloys
Copper (Cu)	Pure metal	• **Malleable** and ductile • Good conductor of heat and electricity • Can be soldered easily • Corrosion resistant	• Electric cables • Plumbing pipes and fittings • Hot water cylinders
Tin (Sn)	Pure metal	• Weak and soft, ductile and malleable • Excellent resistance to corrosion	• Plating of steel food cans • Alloyed with lead to make solder
Lead (Pb)	Pure metal	• Soft and malleable • Excellent resistance to corrosion and radiation	• Insulation screens and barriers against radiation and X-rays in hospitals
Zinc (Zn)	Pure metal	• Very weak and ductile but very difficult to work • Very resistant to corrosion and damp	• Alloyed to make solder • Used in the building trade for flashing joints between structures • Protective coverings for railings and dustbins

Range and applications of pure metals

Non-ferrous alloys

Non-ferrous metals can also be used to make **alloys** in the same way that **ferrous** metals are combined. Brass is the most common non-ferrous alloy and it is made of 65 per cent copper and 35 per cent zinc. Brass is an excellent material to turn on a centre lathe and it also casts easily. It is very resistant to corrosion which is why it is used extensively in the plumbing industry for taps, valves and couplings.

Brass is also a good conductor of heat and electricity. The pins and internal fittings in a conventional domestic three-pin plug are all made from brass for this reason. Brass is also a little tougher than copper which means it is easier to get the pins in and out of the wall socket without them bending too much.

■ Things to do ■

1 Collect a range of metal products. What material have they been made from and why?

2 Make a list of products made from steel and a list of those made from aluminium.

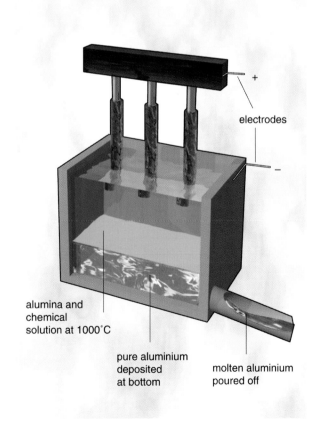

Reduction cell for aluminium – part of the process of electrolysis

Thermoplastics and thermosetting plastics

Aims

- To understand the importance of plastics.
- To understand the range and use of plastics.
- To understand the classification of plastics into thermoplastics and thermosetting plastics.

Most plastics are manufactured apart from a few exceptions which occur naturally, including:

- shellac – an extract from insects to make French polish
- latex – made from rubber, which is extracted from trees
- cellulose – from plants, used to make cellophane for food wrapping and cellulose acetate for cloths and photographic film.

The demand for lighter and stronger materials has led to the continual development of new plastics to replace the more traditional materials such as woods and metals.

Production of plastics

The main source of **synthetic** plastics is crude oil. Oil was formed millions of years ago in shallow seas around land masses. The debris littered on the sea beds eroded and became compacted and as it decomposed, oil was formed. Once the oil has been pumped from the oil rigs to refineries, the production process begins.

The oil is heated up in a **fractioning tower** and is changed into gases. As the gases rise, they are passed through various liquids causing them to separate. The main product which is used in the production of plastics is hydrocarbon naphtha. This is broken down using heat and pressure to form the required molecules.

In plastics, these molecules link up to form large chains of giant molecules. Plastics are divided into two main groups depending upon which way the chains are formed. The groups are:

- **thermoplastics**
- **thermosetting plastics**.

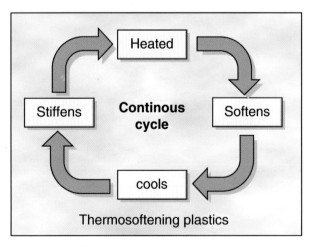

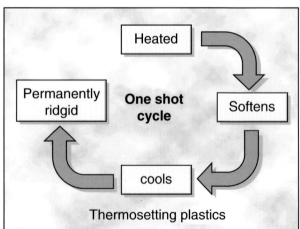

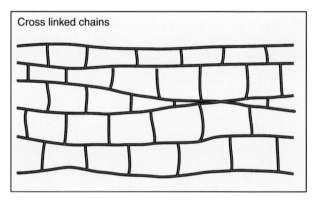

The structure of thermosoftening and thermosetting plastics

Plastic	Type	Properties/characteristics	Uses
Acrylic	Thermoplastic	• Stiff, hard and durable • Easily scratched • Good electrical insulator • Available in a wide range of colours • Polishes and finishes well	• Baths and bathroom furniture • Car indicator covers/reflectors
Polythene (low density) – LDPE	Thermoplastic	• Tough • Resistant to chemicals • Soft and flexible • Good electrical insulator • Available in a wide range of colours	• Squeezy bottles for shampoo and washing-up liquid • Toys • Carrier bags
Polythene (high density) – HDPE	Thermoplastic	• Stiffer and harder than LDPE • Surface has a waxy feel to it • Can be sterilized • Good resistance to corrosion	• Buckets • Bowls • Milk crates • Bleach bottles
ABS	Thermoplastic	• High impact strength • Lightweight and durable • Resistant to chemicals • High quality of surface finish	• Telephones • Kitchenware • Toys
Polyester	Thermosetting plastic	• Stiff, hard and brittle • Very resilient when laminated with GRP (glass reinforced plastic) • Good heat and chemical resistance	• Product cases such as hair dryers • Paperweight castings • Boat hulls with GRP
Epoxy resin	Thermosetting plastic	• Good resistance to wear and chemicals • High strength when used as a bonding agent on **fibrous** materials	• Adhesives • PCB (printed circuit board) material • Lamination of woven sheets such as fibre glass

Properties and uses of thermoplastics and thermosetting plastics

Thermoplastics

Thermoplastics are made up from long chains of molecules which are tangled together and have no formal pattern. There are very few **cross links** between the long chains. This means that when thermoplastics are heated, they become soft which allows them to be bent, pressed or formed into different shapes. As they cool, they become stiff again. The main advantage of thermoplastics is that they can be reheated and reshaped many times. This flexibility is one of the key features of thermoplastics.

Thermoplastics have a 'memory' and when they are reheated they will try to return to their original flat shape, unless they have been damaged by overheating or overstretching. This property is known as plastic memory.

Thermosetting plastics

These type of plastics are made up from long lines of molecules that are cross linked. This results in a very rigid molecular structure. Thermosetting plastics will soften when heated for the first time. This allows them to be shaped, but because they are set in a rigid and permanently stiff molecular structure, they cannot be reheated and reshaped like thermoplastics. The table above gives details of different types of thermoplastics and thermosetting plastics.

■ Things to do ■

1 Using a flowchart, explain the production of plastics.

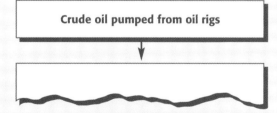

2 Draw up a table and add notes to explain the differences between thermoplastics and thermosetting plastics.

Thermoplastics	Thermosetting plastics

Plastics and carbon fibres

Aims

- To understand the common shapes and available forms of plastics.
- To understand the use of carbon fibre.

Common shapes and forms of plastics

Plastic materials are supplied in a wide range of different forms – powders, granules or pellets, tubes and rods, sheets and film and resinous liquids – depending upon their end use. For example, a powder form would be required for dip-coating, whereas pellets or granules would be best suited for **injection moulding** (see page 52 for a diagram).

Plastic sections and mouldings are all produced by the process of **extrusion** (see page 114 for a diagram). Extrusion involves pushing heated plastic granules through a die to produce an endless length of a uniform cross-section.

Many components, or parts, used in the building industry such as gutters, waste pipes, soil pipes and plastic central heating pipes are produced using the extrusion process.

Carbon fibres

In recent years, carbon fibre has been developed in a similar form to that of glass fibre.

Carbon fibre is a composite. It consists of a matrix, in the form of a plastic resin which sets hard and is reinforced by carbon fibres, the matting. The matting is chopped-up strands woven together. Carbon fibres are woven in a uniform pattern and are very strong. Among other things, they are used for:

- structural components, or parts, in aircraft propellers
- protective clothing for military personnel
- body armour for the police force
- sports equipment such as tennis and squash rackets and skis.

Because of their strength, their use has allowed designers and engineers to reduce the weight of products significantly.

One of the best known uses of carbon fibres was in the production of the 'Lotus superbike' which Chris Boardman rode to Olympic glory in Barcelona in 1992. The single-piece carbon fibre frame replaced

Chris Boardman on his Lotus superbike

All of these items are made of plastic

the conventional tubular steel frame, making it lighter, more streamlined and aerodynamic.

Carbon fibres are available in various forms such as single strands, woven matting or in a woven sock allowing items such as tennis rackets and golf clubs to be made easily.

Although they are very strong, carbon fibres need to be laid up in a mould in a similar way to glass fibre moulds. They are bonded together with polyester type resins to create even stronger lightweight structures.

■ Things to do ■

1 For each of the four products shown in the photographs above, suggest the most appropriate type of plastic.

2 Explain the reasons behind your choice stating what properties the plastic has that make it the most suitable material for that product.

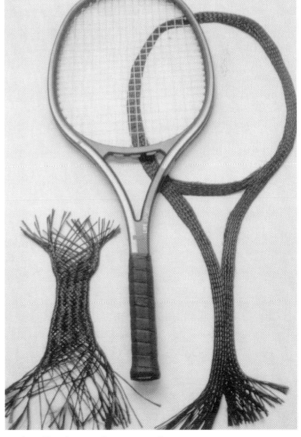

Carbon fibre lay-up of a tennis racket

Classification of components

Aims

- To understand the various adhesives available and their applications.
- To understand the health and safety issues when using specific adhesives.

Fabrication

Most products are made from many different individual pieces of varying materials, all of which need to be joined together in one way or another. **Fabrication** is a general term which is given to this type of assembly.

Fabrication or joining processes can be split into two main categories:

- permanent joints – once made cannot be reversed without causing damage
- temporary joints – not necessarily designed to be taken apart, but they can be disassembled if needed.

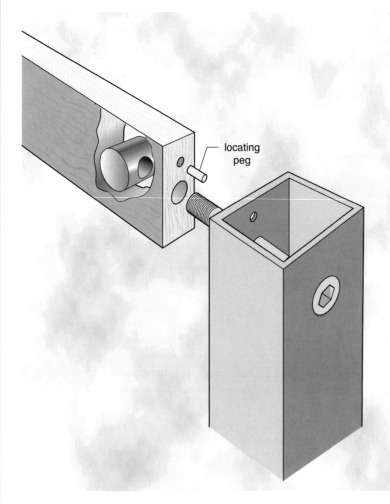

Frame fixing using temporary fixing methods

Adhesives

All joints that use adhesives are permanent. It is very important that the correct adhesive is used for the specific materials being joined.

When using adhesives you should always try to make the glueing or contact area as large as possible.

The five most common adhesives used in the school workshop are:

- polyvinyl acetate (PVA)
- cascamite
- contact adhesives
- epoxy resin (Araldite)
- Tensol cement.

Polyvinyl acetate

This is probably the most common type of wood-working adhesive. It is easy to use and apply and it is very strong, providing that the joints you are sticking form a good fit. The work needs to be held under pressure with clamps while it goes off and any excess should be removed with a damp cloth before it dries. You should remember that most types of PVA are not water-proof. For an external project you must use a waterproof type.

Cascamite

Cascamite is a synthetic resin and is much stronger than PVA. It is also water-proof which makes it suitable for external projects such as garden benches and patio furniture. It is also widely used in the building of wooden boats. It is supplied as a powder and must be mixed with water. Cascamite can be used to fill any small gaps in between any joints. The joint being glued must be held in clamps while the glue hardens and any excess material should be removed with a damp cloth.

Contact adhesive

Contact adhesives, as their name suggests, stick on contact. They are used extensively for glueing large sheet materials such as thin laminates to kitchen worktops. Both surfaces must be coated with a thin layer of the adhesive. It is then left to dry in the air for approximately 15–20 minutes. It is only ready to bond when the surface is touch dry.

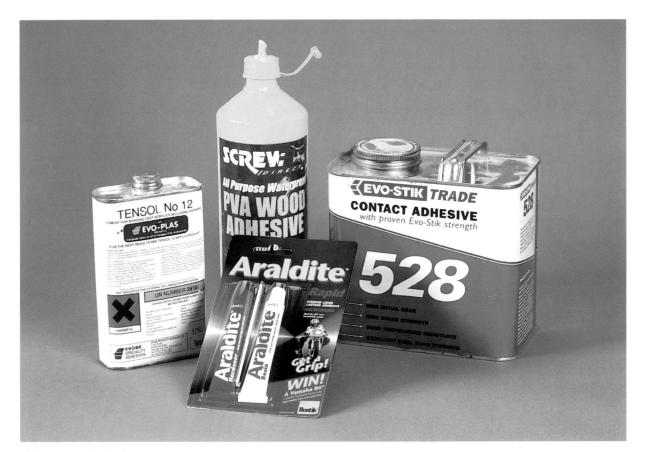

Various types of adhesives

As soon as the two coated surfaces are brought into contact, adhesion takes place and there can be no re-positioning.

Contact adhesive can be used to stick dissimilar materials together such as aluminium sheet on to MDF. Because of the chemicals used in contact adhesive, you should only use it in well-ventilated areas and wear breathing apparatus.

Epoxy resin

Epoxy resins are very versatile adhesives but very expensive. They can be used to bond almost any clean, dry materials. The adhesive is supplied as a two-part adhesive and when required, equal parts of resin and hardener should be mixed to start the chemical reaction. Hardening of the resin starts at once, but full strength of the adhesive is not achieved until two to three days afterwards. Epoxy resins can be used to bond dissimilar materials, but smooth surfaces need to be roughed up slightly. This allows the adhesive to grip better.

Tensol cement

Tensol cement is only used for glueing acrylic. It is a clear liquid with a solvent base that evaporates easily. It must be applied to the joint once put together.

It works by attacking the surface that it has been exposed to. It is not a very strong adhesive and therefore every effort must be made to make the glueing area as large as possible. Any area that you do not want the adhesive to come into contact with should be masked off to avoid any surface damage.

Safety tip

Many adhesives are now solvent-based and give off harmful fumes. Make sure that you use them in areas with plenty of ventilation and avoid contact with skin. Always read and take note of the manufacturer's instructions and warnings.

▪ Things to do ▪

1 Look at a can of Tensol cement or a contact adhesive and draw the symbols pictured on the labels.

2 Find out what each of the symbols means and make a note.

3 Why should adhesives like Tensol cement and contact adhesives be used in well-ventilated areas?

General hardware 1

Aims

- To understand in what form nails are available.
- To understand how nuts and bolts are used.
- To understand how to use screws and how to prepare the holes.
- To understand the use of springs and hinges.

Nails

Nails are a quick method of joining wood. As nails are driven into the wood, they grip by forcing the fibres of the wood away from the head. This then makes it difficult to withdraw the nails.

The length of the nail is important. As a general rule, it should be three times longer than the width of the wood being joined. Nails are generally sold according to their type and length.

Nuts and bolts

Nuts and bolts provide a temporary fixing.

A nut is a collar, usually made from metal, that has a threaded hole through the middle into which a threaded bar or bolt fits. Although generally hexagonal in shape, there are also square nuts and wing nuts. A lock nut has a special nylon insert which stops it coming loose.

Bolts are often made from high **tensile** steel which makes them very strong. They generally have a hexagonal head at one end with a screw thread cut at the other. When using nuts and bolts, a plain washer should be placed between the bolt head and the piece of work and another between the work and the nut. This is to protect the surface from being damaged when the nut or bolt is tightened.

Type	Description
Round wire	• Made from steel wire • Round in section with a flat head • Used for general purpose joinery • 12–150 mm in length
Oval wire	• Made from an oval section wire • Head can be punched below the surface • Used for fixing floorboards to joists and general joinery • 12–150 mm in length
Panel pin	• Thinner pins in lengths up to 50 mm • Small heads can be punched below the surface with a nail punch • Used for finer work such as mitre joints and small lap joints
Masonry	• Made from hardened steel • Round in section • Designed for hammering directly into brick walls

Types of nails

Screws

Wood screws offer a strong and neat method of fixing wood. They can be removed easily and are therefore temporary, unless they are used with an adhesive.

Screws are classified by their length, **gauge**, type of head and material. They are usually made from steel or brass. Brass screws are normally used where steel would rust, such as outside.

It is important to follow these tips when using screws to avoid splitting the wood or damaging the screw head.

1. Screw through the thinnest piece into the thicker one.
2. The screw should be three times as long as the piece being fixed.
3. Drill a clearance hole through the piece being joined slightly bigger than the shank of the screw.
4. Either drill a **pilot hole** or make one with a bradawl.
5. Use a countersink drill bit to make a countersunk recess.
6. If using brass screws in hardwood, use a steel screw of the same size first. Then replace it with a brass screw.

Recently, a new range of screws have come on to the market, such as screws which need no pilot holes, **clearance holes** or **countersink holes.** They can drill themselves in and have serrated edges on the underside of the countersink.

Springs

Springs are used to control movement, apply forces, limit impacts, reduce vibration or measure forces. There are compression or tension springs. Compression springs shorten under a compressive load and their ends are normally ground flat to provide a seating. Tension springs have hooks on their ends and are normally pre-tensioned so that a small load is required to open the coils.

Hinges

Hinges enable a door to move and support it as it opens. The most common type is the butt hinge, widely used in cabinet and furniture construction. One flap, or leaf, is screwed into a recess cut into the side of the door while the other is recessed into the frame. The pivoting section of the hinge projects just beyond the edge of the door and frame when the door is closed.

■ Things to do ■

1. Hammer two oval wire nails into a piece of soft-wood, one with the long axis parallel to the grain and the other across the grain. Observe and record what happens.

2. Disassemble a retractable ballpoint pen and a stapler and look at the springs inside. Are they tension or compression springs? Why?

3. Produce a flowchart to describe the process of preparing and inserting a screw to join two pieces of wood.

4. Draw and label the various parts of a screw, including the head, shank, thread and its gauge size.

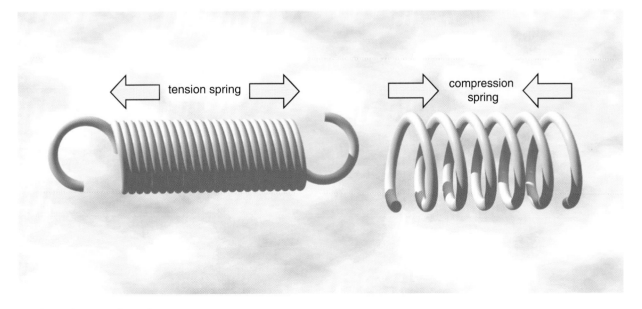

Tension and compression springs

General hardware 2

Aims
- To understand the range of catches and handles available.
- To understand the use of knock-down fittings and structural items for fixing.

Catches and handles
There is an almost endless supply of various catches and handles. For catches, ease of use and accessibility are important. For handles, careful consideration must be given to their weight and how easy the user will find them to grip or grasp.

Knock-down fittings
There are many modern knock-down jointing methods and they all allow strong joints to be made quickly and easily. The parts can also easily and quickly be taken apart, so that the whole construction can be 'knocked down' or flat-packed for easy transportation or storage.

The most common form of knock-down fittings are nylon joint blocks screwed into the corners, one against each face. They are then fastened together by a steel bolt which locks the two boards to each other. There are many variations on these such as assembly joints, a single block and a moulded single piece with a flap to conceal the screw heads.

Structural items
The range of general hardware goes well beyond what has been looked at here. Many structural items also exist in the form of fixing plates and braces, shelf brackets and supports and wall fixings.

Shelving systems
Shelving systems are available as either fixed or adjustable. The adjustable type allows you to move the shelf brackets up and down depending upon the height of the items on the shelf, for example books. Fixed shelves, as their name suggests, are permanently fixed in one position and cannot be moved.

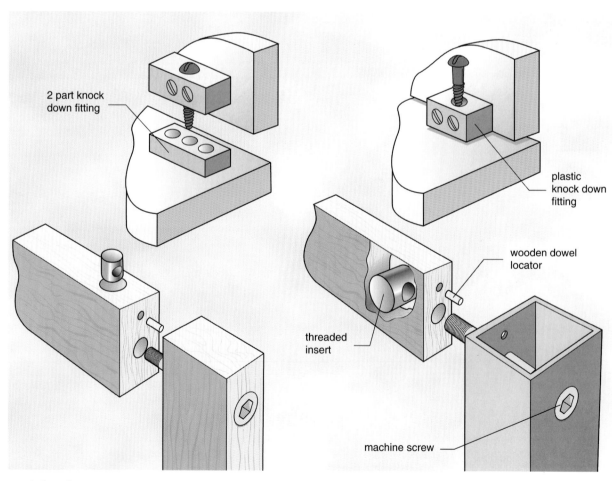

2 part knock down fitting

plastic knock down fitting

wooden dowel locator

threaded insert

machine screw

Knock-down fittings

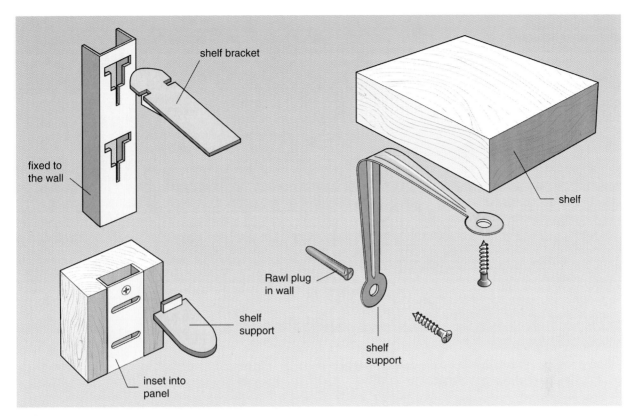

Shelving systems

Rawl plugs

Rawl plugs must be used when fixing the brackets to the walls. They are only suitable for use in solid walls. The appropriate size hole is drilled and the rawl plug inserted. When the screw is inserted and screwed in, the rawl plug expands and grips against the side of the hole providing a very strong means of anchoring the bracket to the wall.

Angle and tee plates

Angle and tee plates are pressed from sheets of steel and are used to create either right angle joints or tee joints.

▪ Things to do ▪

1 Disassemble/assemble a piece of flat-pack furniture such as a video cabinet. Look at the instructions and see how easy they are to follow.

   a Make a sketch of the different types of knock-down fittings that have been used in the construction of the product.

   b List the different components that have also been used such as panel pins and screws.

   c Make a list of the various materials that have been used for the actual product itself.

   d Explain why they have used these materials.

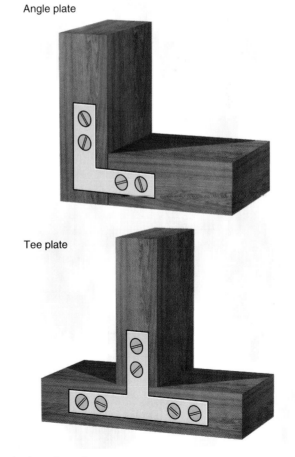

Angle and tee plates

Properties and working characteristics of materials 1

Aim

- To understand the aesthetic, physical and mechanical properties of materials.

Choosing suitable materials

Engineers and designers need to be aware of the properties and limitations of the materials they intend to use and to take into account that all products will be associated with or subjected to a force. There are two types of properties of materials – **mechanical properties** and **physical properties**.

Mechanical properties

These refer to the behaviour of materials when subjected to a force. For example:

- compression
- tension
- hardness
- durability
- toughness
- elasticity
- plasticity

Physical properties

Properties not associated with the application of force. For example:

- density
- corrosion resistance
- electrical resistance
- fusibility
- ability to reflect light
- thermal properties

See page 27 for a more comprehensive list.

The choice and suitability of metals for specific products offers a wide range of alternatives, but the material must be suitable for the purpose and use of the product. For example, although steel is available in many alloyed forms, stainless steel would be an inappropriate choice of metal for a fizzy drinks can because of the quantity required, the cost and the difficulties in actually manufacturing the can.

Similarly, mild steel would not be the best material from which to make knives and forks because, although the manufacturing processes are available and cost is not an issue, mild steel **oxidizes** in air and forms a coating of iron **oxide**, a red dusty powder. It would simply be unhygienic to use.

A kitchen saucepan must display many different properties, both physical and mechanical. It must be made from a hard tough material in order to stand up to knocks and bangs. It must also be lightweight and a good conductor of heat, but be able to withstand high temperatures. It also needs a handle that is made from a poor conductor of heat, so that it doesn't get too hot. We therefore end up with a number of different materials being used in a single product because of their different properties.

The best choice of material depends on many factors:

- its properties
- its **aesthetic** qualities
- the cost of both material and production
- its availability.

All of these factors must be taken into account when designing a product. A change in one area will have a knock-on effect on another.

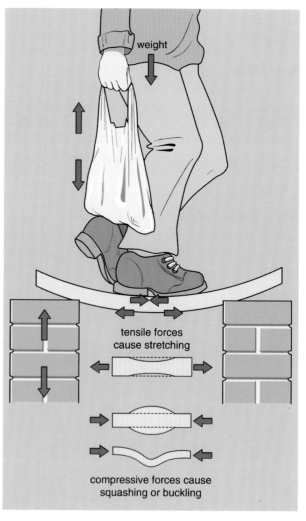

Mechanical properties at work – both tensile and compressive stresses develop in the beam when weight is added.

Woods

Wood is a natural material and factors such as soil conditions and the climate affect its growth. Therefore, we can only give general details about its properties as even pieces of wood taken from the same tree may have different characteristics and properties.

The general appearance of wood grain, colour and texture is probably the most important consideration when choosing the type of wood to use. The working characteristics and strength are usually secondary considerations, but they are still important. A balance needs to be struck between appearance, strength, workability, cost, weight and availability.

The decorative finish of hardwoods is generally superior to the softwoods. The main characteristics are determined by the cell structure which gives rise to the grain, figure and texture:

- Grain – the mass of a wood's cell structure represents the 'grain' of the wood. The grain of wood follows the major direction of growth, which is vertically up through the tree.
- Figure – the figure of wood is the difference in growth between early and late wood, the density of annual rings, the distribution of colour and even the effect of disease or physical damage. The figure of a wood is often best shown when it has been cut in a particular way at the sawmill. Oak is particularly well known as having a good 'figure'.
- Texture – this refers to relative sizes of the cells. It is also used to describe the distribution of cells in relation to annual growth rings.

An antique stationery box with inlaid veneers set into the top

Wood and wooden products are not really associated with surface decoration other than simply to enhance and show off the natural grain and beauty of the timber.

Techniques such as marquetry and parquetry can be used to decorate surfaces. This is where thin layers of veneers are laid into the solid top to form motifs, pictures or patterns.

The terms hardwood and softwood are not descriptions of the wood's mechanical properties. Generally, hardwoods tend to be stiffer than softwoods because of their fibre content. Some timbers such as ash display better properties of toughness and elasticity than others. This makes it an ideal timber for sports equipment and tool handles. It is also used widely where timber needs to be bent or curved.

Timbers have fairly good **tensile** strength, the ability to resist being pulled apart. However, the compressive strength of wood is quite weak. It is worth noting that wood is very weak in both tension and compression across the grain.

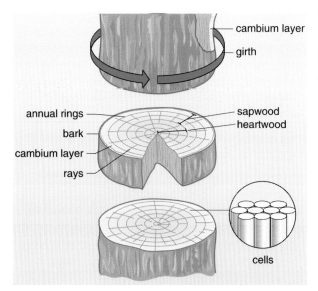

▪ Things to do ▪

1. Collect some samples of the following timbers: oak, mahogany, beech, ash, birch, pine. Look carefully at the grain, figure and texture and record your observations using annotated sketches.
2. Try slightly flexing each of the samples along the length of the grain and make some comparisons about how elastic they are.

Properties and working characteristics of materials 2

Aim

● To understand the aesthetic, physical and mechanical properties of plastics and metals.

Plastic is the best material for a bleach bottle

Plastics

The term plastic describes a material which at some point behaves in a putty-like way. It will deform under pressure and retain the new shape as it cools down and the pressure is removed. All plastics display good resistance to corrosion and all have a low density relative to the majority of woods and metals.

However, both groups of plastics – **thermoplastics** and **thermosetting plastics** – demonstrate particular characteristics. Thermoplastics are easily moulded and any waste can be re-used. However, they are less useful where heat is concerned as many thermoplastics soften and lose their rigidity at temperatures just above 100°C. However, thermosetting plastics are much better at retaining their rigidity at higher temperatures. They also make good thermal insulators.

Due to the nature of the material, plastics are available in an almost endless range of colours.

Surface texture is applied to the product being made in the manufacturing stages. The required texture forms part of the mould being used and as the plastic is forced into the mould, it takes the shape and form of the mould including any texture on the mould. One major benefit of this has been to mould braille on to the surface of products such as bleach bottles.

Plastics are good electrical insulators and therefore provide the ideal material to coat electrical cable in. Low density polythene is softer, weaker and more flexible than the high density type. It is therefore a suitable material with which to insulate copper cables because they need to be flexible when being pulled through holes and in between joists (NB: PVC is normally used for wire and cable insulation). High density polythene is much stiffer and stronger and it is also quite tough. It is an ideal material from which to make products such as watering cans and water butts. Once moulded, it will retain its shape and stiffness and will withstand sudden knocks and bumps without splitting or breaking.

Metals

The appearance of metals is generally cold and the only texture to be considered is that which can be moulded or machined into the surface.

Colour is often applied by painting or dip-coating and it also serves to protect the metal surface from oxidizing or corroding. Therefore, surface finish is both **aesthetic** (it looks good) and functional (it protects the surface).

In general, all metals are good conductors of heat and electricity. They are also generally mechanically stronger than most woods and plastics.

The table opposite gives some examples of materials and their **physical** and **mechanical properties**.

The selection and choice of a material or materials for a product or component, or part, is not an easy one. Background knowledge about properties, available forms and sections, manufacturing processes and aesthetic qualities of materials can all influence and justify the selection made.

■ Things to do ■

1 What plastic is best suited for use as a bleach bottle?

2 Why is it a good idea to mould braille on to the surface of products such as bleach bottles?

3 The tips of soldering irons are made from copper. What property does copper have that makes it an ideal material for this application?

4 Explain why mild steel is used for car body panels rather than aluminium.

5 Explain and describe what properties acrylic has which makes it suitable for use as a bath.

Property	Physical/ mechanical	Description	Material	Applications
Electrical conductivity	Physical	Low resistance to the flow of an electric current	Gold, silver, copper and aluminium	Electrical cable
Thermal conductivity	Physical	Measure of how much heat travels through a material	Metals generally, but copper especially	Car radiators
Thermal insulator	Physical	Materials which have a low value of thermal conductivity	Non-metals generally	Cavity wall insulation
Fusibility	Physical	Ability to change into a liquid or molten state at a certain temperature	Tin Lead Aluminium	Solder Castings Alloys
Density	Physical	The mass per unit volume kg/m^3	High – lead Low – expanded polystyrene	Fishing weights Buoyancy aids
Optical	Physical	Materials react to light in different ways – reflection and absorption	Wood is opaque Glass is transparent	Acrylic reflector lenses for cars
Strength	Mechanical	There are five types: • tensile • compressive • bending • shear • torsion	Materials behave in different ways when subjected to different forces	Grades of steel have been developed to resist these different types of forces
Elasticity	Mechanical	Ability to return to original shape after force has been removed	Spring steel Rubber Ash	Springs Sports racket handles
Plasticity	Mechanical	Ability to be changed permanently without cracking or breaking	Acrylic ABS	Baths Plastic moulded products
Ductility	Mechanical	Ability to be drawn or stretched	Silver Copper	Electrical cables
Malleability	Mechanical	Ability to be deformed by compression without tearing or cracking	Lead	Roof flashing
Hardness	Mechanical	Ability to withstand abrasive wear and indentation	High carbon steel Silver steel	Drills Files Scribers
Toughness	Mechanical	Ability to withstand sudden shock loading without fracture	Mild steel	Nails Screws Car body panels
Durability	Mechanical/ physical	Ability to withstand weathering deterioration and corrosion	Plastics generally Gold	Window frames Jewellery

Physical and mechanical properties of materials

The choice and fitness-for-purpose of materials 1

Aim

- To understand that the choice of material depends upon a number of factors.

The selection and choice of a material or materials for a product or component, or part, is not an easy one. Background knowledge about properties, available forms and sections, manufacturing processes and aesthetic qualities of materials can all influence and justify the selection made.

The final factor relates to the scale of production and therefore intended processes of manufacture. It would not be viable to **injection mould** a single one-off product because of the very expensive setting up and tooling costs. On the other hand, you would not expect to hand cut 10 000 jigsaw puzzles to be sold in a high street chain of toy shops.

Some examples of materials chosen for fitness-for-purpose
A hanging basket

The hanging basket bracket pictured below left is made of mild steel. The manufacturing processes used to produce it are fairly simple:

- The materials are cropped/cut to length.
- Holes are punched in the bracket allowing it to be fixed to the wall.
- Scroll work is formed on each piece.
- The separate components are assembled in a jig.
- The components are welded together.
- The assembled project is sprayed or dip coated.

Mild steel is widely available in a variety of sections and 12 mm x 3 mm is a suitable sized section for this purpose. It is easily cut on a guillotine and the holes can also be punched quickly.

Mild steel is sufficiently ductile and malleable to allow for the scrolls to be cold formed by hand. Mild steel can also be welded easily. The bracket is likely to either spot or **MIG welded**. Both of these welding processes are quick and effective methods providing a very strong joint.

Mild steel is also strong enough in compression to withstand forces acting upon it with the weight of the basket (See the diagram on the next page).

If the steel was not protected in any way, an **oxidized** film would soon form on the surface, and the bracket would start to rust. Spray painting or dip coating are the two most common methods of finishing this sort of product.

Bracket made of cheaper flat mild steel

Ornate bracket made of cast iron

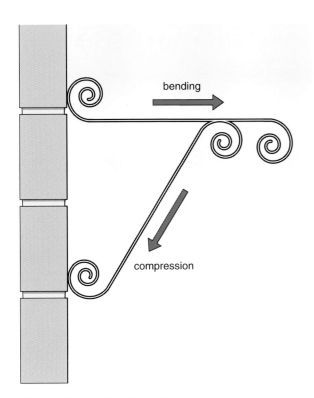

Forces acting on a hanging bracket

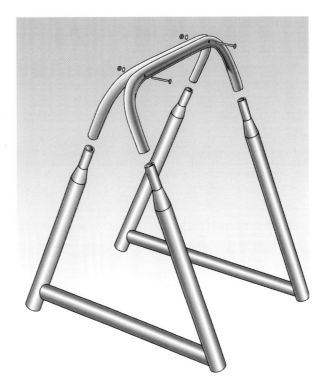

Exploded view of child's swing

A garden swing

The choice of material for a child's garden swing may conjure up an image of an immensely strong, well-engineered structure. However, many garden swings are made from lightweight round tubular sections that are simply bolted together using standard nuts and bolts. In order to stop the nuts from coming loose as a result of the vibrations, a nylon lock nut is used.

The selection of tube may be somewhat surprising. However, the forces generated as a result of the swinging motion are countered by the wide stable base and the triangulated nature of the frame. The horizontal bar forms a triangle with the two diagonal legs, making a very rigid structure.

The frame itself is subject to mainly bending with a degree of tension. Of all the sections commercially available, round section tube is the stiffest and quite resistant to bending. The combination of both of these properties makes the choice of round tube the most appropriate for the task.

The steel frame is often galvanized, a process whereby a layer of zinc is deposited on to the surface, stopping it from oxidizing and then rusting.

The top two rails have a right-angle bend formed at each end. This forming process is easily carried out in the school workshop and in industry. In school workshops, pipe benders are used and the mechanical nature of steel allows the cold working to be carried out easily.

The seat is often injection moulded from a thermosetting plastic such as polypropylene. It is a rigid material and is also very tough. However, if left exposed to direct sunlight, its colour fades which results in a rather unattractive appearance.

The combination of all these individual components and structural members, once assembled, results in a substantial, safe product. One major advantage for the consumer is that it can come in flat-pack form. This means that it can easily be transported from the shop and assembled at home.

■ Things to do ■

1 The mild steel bracket shown opposite is just one manufacturing solution to the problem of hanging up the planted basket. A second cast solution is also pictured, but it has a retail value of almost five times as much as the mild steel version.

Find out about the process of casting and use this to explain why the two products are priced so differently.

The choice and fitness-for-purpose of materials 2

Aim

- To understand that the choice of materials depends on the relationship between working properties.

Further examples of materials chosen for fitness-for-purpose

A cool box

Physical properties are very much in evidence when it comes to selecting materials for use in a picnic cool box.

The material obviously needs to be as lightweight as possible since the cool box is portable. The material should also be reasonably tough due to the knocks and bumps it is likely to receive. The main purpose remains, however, to keep the contents of the box cool. It is for this reason that most of the research and investigation should be focused around the thermal properties of the materials available.

In this case, the designer would need to consider good insulators of heat. Even though we want to keep the contents cold, we need to stop the box being warmed up by the surrounding air temperature. Cork, expanded polystyrene, glass fibre and paper, are all very good insulators of heat. They are very effective

because they all make use of pockets of trapped air inside the material and air is one of the best insulators of heat.

The main structure of a cool box is usually **injection moulded** from a thermoplastic material. This provides a stiff, rigid, tough exterior and interior casing. The gap between the two skins is filled with either a foam or expanded polystyrene. Both of these materials are low density and therefore do not weigh very much in relation to the volume (amount) of material used. They also contribute to the overall stiffness of the product as they help to reinforce the two plastic shells.

A child's toy

Beech is a particularly good hard wood for making children's toys. Its grain pattern and texture are **aesthetically** pleasing, especially when finished in a clear varnish or lacquer. It does not splinter easily and it is quite hard wearing although it does blunt sharp edges on cutting tools easily.

Beech is widely available in plank form and it can be cut and planed to almost any size in a good timber yard. It is also quite commonly found in most school workshops and it has only taken a few very small bits to make the toy shown.

The two side runners were very carefully marked out and clamped together. They were drilled as a pair so

A rigid plastic cool box

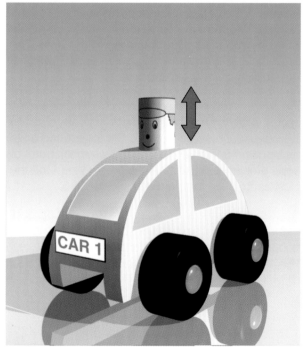

A child's toy – the head moves up and down as the car moves.

Cam fixed on to the front axle

that when the axles were inserted, they would line up and they would be parallel to each other. This marking out process and drilling requires some attention in order to achieve a high level of accuracy.

The wheels were carefully chosen to ensure that they matched up with the available stock sizes of metal rods being used for the axles.

The axles were made from BDMS. It was crucial that the ends were cut square and a small chamfer was filed around the edge. This helps with the fitting of the press caps which are used to hold the wheels on to the axle.

The wheels were allowed to rotate freely on the axle. It would be in this area in which the product would be very carefully tested along with the toxic nature of any surface treatment that might be used.

The toy would be subjected to many different British and international standard tests. It is essential that no small parts can be removed since they might be swallowed by the child resulting in choking accidents. Once the toy has passed all the necessary and appropriate tests, it would be awarded a certificate to indicate that it has passed and all toys of this model would carry a mark to indicate that it is a safe toy.

■ Things to do ■

1   Without being too concerned about the manufacturing processes and techniques, study carefully the materials that have been used in:

    a   a 3-pin plug

    b   an automatic jug kettle

    c   wooden salt and pepper mills.

2   Copy out the table below and fill in the appropriate details about the products.

	3-pin plug	Kettle	Salt and pepper mills
Properties			
Aesthetic qualities			
Materials used			
Manufacturing processes involved			

1 The drawing below shows a type of wood joint marked out ready to be cut.

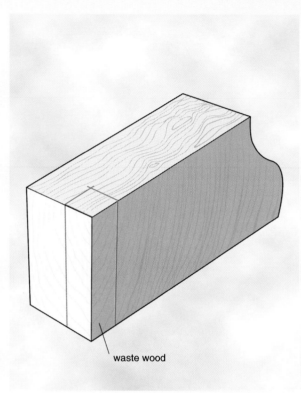

waste wood

A wood joint marked out

a Add to the table below the tools used for marking out this piece of wood. One tool has been named for you. **(2 marks)**

Process	Tool used
Marking out the depth of waste to be removed	Marking gauge
Scoring lines across the grain	
Generating a square edge across the face side	

b Name the two hand tools used to remove the waste. **(2 marks)**

c Make a clear sketch of a marking gauge and identify its key features. **(5 marks)**

d Explain how the marking gauge would be used to mark out the depth of the housing. **(3 marks)**

2 a Explain the meaning of the term 'alloy'. **(2 marks)**

b Name the *two* main materials that are used to produce steel. **(2 marks)**

c Before using a centre lathe, a risk assessment should be done. Describe three precautions you would take before turning the component on the lathe. **(3 marks)**

3 Two types of screw head are shown below.

a Name each type of screw head pictured. **(2 marks)**

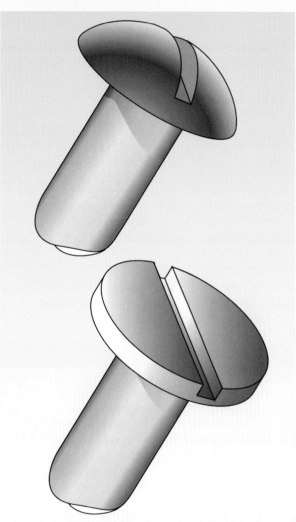

Two types of screw head

b Explain why a steel screw is inserted into hardwoods before a brass screw can be used. **(2 marks)**

Section B
Preparing, processing and finishing materials

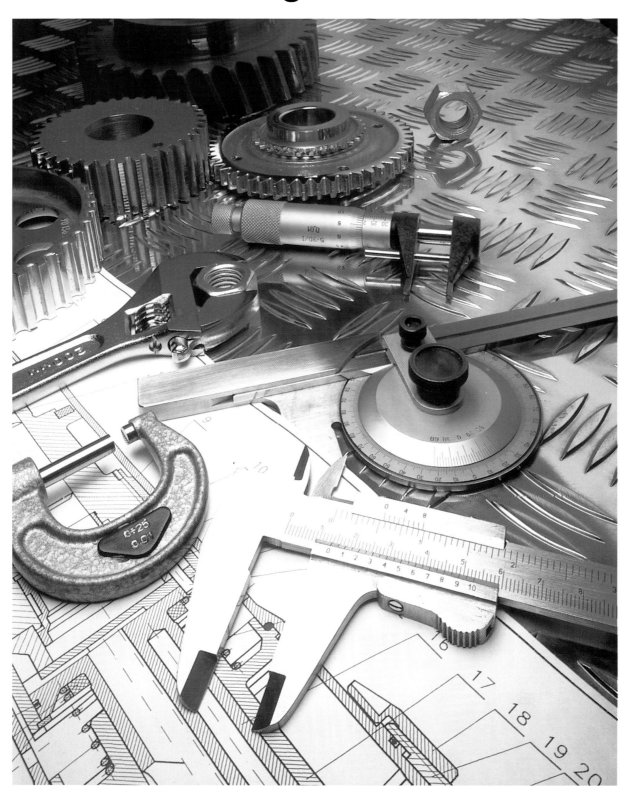

Heat treatment processes

Aims

- To understand the heat treatment of metals.
- To understand the processes of annealing, normalizing, hardening, tempering and case-hardening.

Heat treatment

The term heat treatment describes the process of heating and cooling metals in a controlled manner. This allows us to change the properties of the metal, such as increasing the **hardness** or reducing the brittleness. Heat treatment processes can be broken down into five main areas:

- annealing
- normalizing
- hardening
- tempering
- case-hardening.

Annealing

As a metal is worked or deformed by bending, rolling, or hammering, its structure changes. As a result of this cold working, the material's hardness increases. This makes it increasingly more difficult to work. To ease this, the material needs to be annealed. The annealing process will restore the initial structure of the material by relieving the internal stresses.

The process of annealing involves heating up the metal to a certain temperature and then allowing it to cool. It is quite common practice to dip or 'pickle' cooled brass and copper in dilute sulphuric acid. This process chemically removes the scale which formed on the surface during the annealing process.

Normalizing

Normalizing is a process that only applies to steel. The grain structure in steel becomes coarse as a result of work hardening. The refining of the grain restores ductility and toughness.

The normalizing process involves heating the material to a temperature of 700–900°C depending upon its carbon content. The work piece should be held at this temperature for a short while, normally called soaking, before being allowed to cool in still air.

Forged components such as hooks for cranes or kitchen knives undergo the process of normalizing to restore their **mechanical properties**.

Hardening

It is only possible to increase the hardness of steel that has more than 0.4 per cent of carbon in it. The full effects of hardening will not be possible unless the carbon content is above 0.8 per cent. In order for steel to be used to make tools such as scribers, drills and punches, it must be possible to increase their hardness. Therefore they must be made from medium carbon steel or silver steel.

In order to fully harden a piece of silver steel, it should be heated to just over 720°C. It should be soaked at this temperature until it is uniformly heated and then quenched immediately into water.

At this point, the component is very hard, but is too brittle to be of any practical use. The hardness needs to be slightly reduced to produce a more elastic, tougher material that will retain a **cutting edge**. This is done through tempering.

Approx. temperature (°C)	Colour	Toughness	Uses
230	Pale straw	least	Lathe tools
240	Straw		Scribers
250	Dark straw		Centre punch
260	Brown		Tin snips
270	Brown–purple		Scissors
280	Purple		Saw blades
290	Dark purple		Screw driver
300	Blue	most	Springs

Tempering colour chart

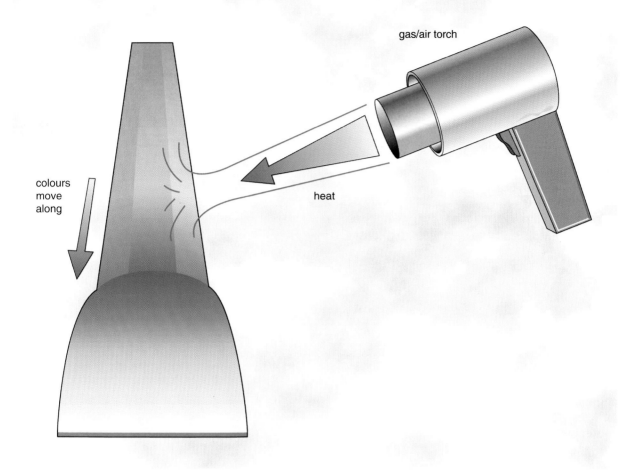

gas/air torch

heat

colours move along

The tempering of a cold chisel

Tempering

To temper a piece of hardened steel, it must first be cleaned with emery cloth or wire wool. This cleaning allows the colour of the **oxide** formed on the surface to be easily and clearly seen as the component is reheated.

The steel piece needs to be heated well behind the cutting edge. As the metal heats up, an oxide appears on the surface and this oxide can be seen to pass along the metal as it gets hotter.

When the appropriate colour reaches the cutting edge or tip, it should be quenched (cooled) immediately in water. The hardness and brittleness are both reduced as the temperature increases.

Case-hardening

The only way to harden mild steel that does not contain sufficient carbon is to case-harden it. This gives it a carbon-rich skin while keeping the more elastic, ductile properties on the inside.

The metal must be heated to a cherry red and then dipped into a carbon powder. It should be left there to cool. This process should be repeated three or four times, each time increasing the hardness.

The final step is to heat again to a cherry red colour and then to quench in water. This process is useful for producing tool holders and driving shafts that require a hand outer shaft.

■ Things to do ■

1 Harden and temper a piece of silver steel. Observe the coloured oxides on the surface when tempering. Make sure you follow all safety precautions.

2 Draw up a table using the headings annealing, normalizing, hardening, tempering and case-hardening. Add definitions of each of these terms to the table.

3 When making a screwdriver blade, why must it be both hardened and tempered?

4 Produce a flowchart (with illustrations) for the process of case-hardening.

Alloying and laminating

Aims

- To understand what an alloy is and why it is used.
- To understand the use of laminated materials to improve properties.

Alloying of metal

The process of alloying metals has allowed new materials to be developed. These new materials, or alloys, often have improved properties.

Brass

Brass is an alloy of copper and zinc.

Zinc as a pure metal is weak and difficult to work. However, when alloyed with copper to form brass, it displays greater hardness and it becomes much easier to work. It keeps its resistance to corrosion which makes brass an excellent material for use as boat fittings and valves.

High speed steel

High speed steel is an alloy of medium carbon steel with tungsten, chromium and vanadium.

It is an exceptionally hard material which can only be ground. It is very resistant to heat generated as a result of friction which makes it an ideal choice of material for lathe cutting tools, milling cutters and drills.

Solder

Solder is an alloy of lead and tin.

The process of **soldering** can be used to join copper, brass and tin plate. Solder has a very low melting point. Solder contains more tin than lead. This makes for a quicker 'freezing' time.

Laminating

Laminating is a process that can be applied to:

- woods
- metals
- plastics.

A laminate is a single piece or sheet of material and when several pieces are stuck or bonded together they are said to have been laminated (see the diagram on page 48).

Laminating wood

Laminating wood allows **mechanical properties** such as strength to be increased. Single wooden laminates are bonded together in a flat form to produce plywood. Thin laminates can be stuck together and trapped between a former to produce curved shapes such as chair backs or legs. In a school workshop clamps are generally used to apply the pressure but in industry vacuum tables are normally used.

Strips of timber can also be bonded together to form big sections of wood. These are then used as structural supports. They have great **mechanical strength** and stability that can resist large bending forces. Timber is often used like this since it offers a very attractive solution if it is simply varnished allowing its natural grain and colour to be seen.

Laminating metals

Dissimilar materials such as brass and **steel** are laminated to form a bi-metallic strip. These two separate metals have different co-efficients of expansion, which means that one moves more than the other when heated. Since they are joined together, they curl as they get hotter. Bi-metallic strips are used in kettles and heating control mechanisms as switches.

Laminating plastics

Plastics can be laminated in either a flat-sheet form to produce hard wearing chemical resistant surfaces, or together with glass strand matting and epoxy resins to form curved shapes. Plastics are generally used in the form of thermosetting resins with fibres embedded within them. Great mechanical strength can then be achieved with a range of surface finishes and colours available.

Plastic memory

Thermoplastics are plastics that can be reheated and reformed, such as acrylic. Because of this,

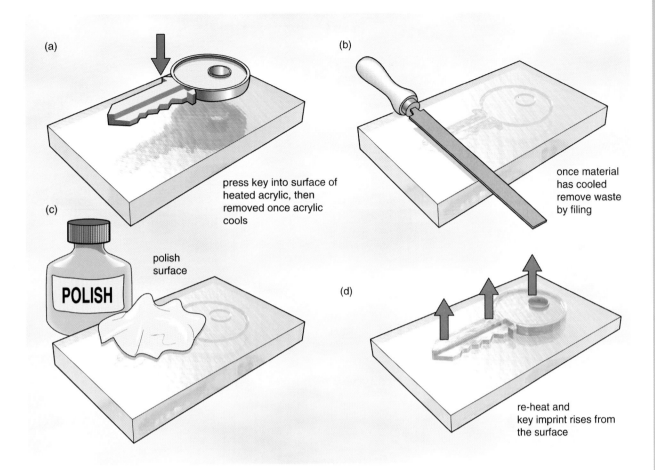

(a) press key into surface of heated acrylic, then removed once acrylic cools

(b) once material has cooled remove waste by filing

(c) polish surface

POLISH

(d) re-heat and key imprint rises from the surface

Experiment to show acrylic's plastic memory

some thermoplastics are said to have a 'plastic memory'. If a mistake is made in bending or forming, they can be reheated and the mistake corrected.

A good example of acrylic's plastic memory is shown in the diagram above. A small sample of acrylic is heated to about 160°C and a key or similar object is pressed into the surface and held there until the material has fully cooled. The remaining material is removed to leave a thinner section which can then be finished and polished. Once reheated, the compressed section will return to its original state which means that the key imprint will rise out of the material.

Composites

A composite material is a material made from more than one substance. Some examples of composite materials are:

- MDF
- GRP (glass reinforced plastics)
- carbon fibres.

Composites are generally regarded as having improved properties. Certainly, both GRP and carbon fibres have good tensile strength, good resistance to corrosion combined with a low density. Carbon fibres in particular have much greater strength than GRP.

The process of using GRP and carbon fibres is quite similar. The two components that are used are the fibrous material and a resin in which the fibres are embedded. As the resin is absorbed by the fibres and the resin cures or goes off, a very strong material is formed. This may be used to form either a very rigid structure such as a boat hull or body armour or a very flexible elastic structure such as a carbon fibre fishing rod or a tennis racket.

■ Things to do ■

1. Carry out the plastic memory experiment, using a key and a piece of acrylic, as shown on these pages.

2. Explain why timber is laminated to improve its strength.

3. What is a composite material? Give three examples.

Finishing processes 1

Aims

- To understand the need for finishing processes.
- To understand the need for thorough surface preparation.

A wide variety of finishing processes and techniques is available for woods, metals and plastics. Generally, plastics do not require a great deal of treatment and surface finishing because of the nature of the material and the manufacturing techniques involved.

Reasons for finishing and surface treatment

Finishing and surface treatment is usually carried out to achieve one or both of the following reasons:

- **aesthetics** – to improve the nature of the material
- function – to protect the material and to stop it from deteriorating and to prolong its useful life.

Suitability of finishing and surfaces treatment

The suitability of a finishing process depends greatly on:

- what material has been used
- where it is going to be used.

You should consider the choice of finish in the early stages of a design and not simply leave it until the end. You also need to consider the durability of the surface finish and the aftercare required. Maintenance is essential if a product or component is expected to last for many years.

When to apply finishes

Generally, finishes are applied before final assembly. This is especially important on any internal surfaces. However, it is a good idea to make sure that any joints that are being glued are taped and covered to avoid the application of a finish.

Preparing surfaces

Before any finish can be applied, it is important to prepare the surface. As already stated, plastics do not often undergo major surface treatments. Woods and metals, however, do need to have their surfaces prepared.

Ferrous metals

In particular, ferrous metals must undergo thorough preparation regardless of the finish. All surface oxides must be removed by emery cloth or wet and dry paper. If they are then going to be painted, they must be degreased. This is done by cleaning the surface with methylated spirits on a rag.

Wood

Wood must also be prepared carefully before any finish can be applied. A plane can be used to produce a clean smooth surface. Any minor surface blemishes should be removed with progressively finer grades of glass paper. It is important to work along the grain to prevent any scratching.

Large surfaces

On larger surfaces, electrical sanders such as a belt or an orbital sander can be used. It is important that these electrical sanders are fitted with dust bags to contain the fine particles.

In industry, large flat belt sanders are used to prepare and finish large flat surfaces such as doors or table-tops. These sanders have large extraction systems and users have to wear ear protection.

Polishes

The two most common forms of polish are wax polish and French polish. Both types of polish bring out the grain of timber.

Wax polish

Wax produces a dull gloss shine. It is made from bees wax dissolved in turpentine to form a paste. It is applied to timber using a cloth. With the addition of a silicon wax or carnauba wax, the durability of the finish is greatly increased.

Before any wax can be applied, the surface of the timber must first be sealed. This can be done using shellac which is a natural resinous product dissolved in either cellulose or methylated spirits. This penetrates and seals the surface.

A wax finish would be used on a wooden desk, coffee table or dining room table where you want to show off the natural grain and colour of the timber.

Stained wooden toy

French polish

French polish could also be used in any of the situations above. It does, however, produce a much higher glossed surface and you are also able to change the colour slightly.

The quality of the finish is dependent upon how it has been applied and how many layers there are. A harder finish is built up with successive layers of shellac which has been dissolved in methylated spirits. The final coat is simply methylated spirits. This creates the high gloss finish.

Staining

Staining or colouring is used to heighten the natural grain of timber. It is very much a decorative finish and allows for an even application of colour.

Stains can be water-based, spirit-based or oil-based. Spirit-based stains tend to dry quicker, whereas oil-based stains last longer and are more versatile.

Stains are available in a wide range of colours and can also be used to create a fake effect. Colours such as oak and mahogany are available to use on cheaper soft-woods to give an impression of a more expensive product.

Stains can be either brushed on or applied with a cloth. In industry, products are either sprayed or dipped allowing for a much quicker and more even application.

Varnish

Synthetic resins (plastic varnishes) produce a much harder, tougher surface. They are heat-proof and water-proof and quite good at standing up to tough knocks.

Many types now exist on the market in a wide range of shades and finishes such as matt or gloss. They are best applied in thin coats with a brush or spray. In between each layer, the previous one should be gently rubbed down with wire wool.

Varnish should always be applied in the direction of the grain in light, even strokes.

■ Things to do ■

1 Draw up a table for the range of finishes and list the advantages and disadvantages of each.

2 For each of the finishes listed above, give two examples of where they might be appropriately used.

Finishing processes 2

Aim
- To further understand the range of surface finishes available.

Paint

Paint can be applied to both woods and metals. It is used to provide a decorative colouring and protective layer whether used indoors or outside.

Painting wood

In preparing wood to be painted, any knots should be sealed with shellac to prevent any resin from seeping through and spoiling the appearance. Any sharp corners should be gently rounded off with some glass paper.

The wood should then be sealed with a primer. An undercoat can be applied after a gentle rub down. Finally, a top coat can be applied. Most paints for wood are either oil based or spirit based. Both of these types are durable and waterproof but the polyurethane type is generally much tougher.

Water based paints are being used more and more, as they produce less fumes (respiratory hazard) and therefore odour, when drying.

Painting metal

The preparation of metals before painting is very important. Surface **oxides** must be removed and surfaces degreased. A red-oxide paint is generally used as a first coat to prevent further oxidation of the surface.

A primer and undercoat are then applied before the final top coat. In between each layer the surface should be gently rubbed down with some very fine wet and dry paper.

Metals are normally sprayed in industry in water chambers to take away the smell and waste material. The spraying of cars in the motor industry is now widely carried out by robots. They have been programmed to follow the shape and contours of the car very carefully and have also given rise to a much better quality of finish.

'Hammerite' is a type of metal paint that doesn't need extensive surface preparation. It can be painted straight on to metal almost regardless of the surface condition. It is typically used on wrought iron gates, fences and old-fashioned cast iron drain pipes. It is also often used on workshop machinery and it can easily be detected due to its 'cracked' textured surface appearance.

Plastic dip coating

Dip coating is a process that is suitable for most metals. It is used for coating metal products such as hanging baskets, brackets, kitchen drainers and tool handles.

The metal must first be thoroughly cleaned and degreased before being heated in an oven to 180°C. It must be soaked at this temperature before being plunged quickly into a bath of a fluidized powder. It should be left there for a few seconds while the powder sticks to the hot surface to form a thin coating. The self heat of the article then melts the plastic powder to leave a smooth glossy finish.

Preservatives

Garden sheds, fence panels and commercially produced furniture are all treated with wood preservatives, such as creosote – a tar/oil-based product – which soaks deep into the surface where it forms a barrier against damp and the entry of water. In addition, timber used in the construction industry for joists and root trusses is treated.

The timber is often 'tanalized' which means that it has been treated under pressure. This results in the preservative soaking deep into the surface. This type of pressure treated timber is especially effective against damp.

Etching

Etching is a finishing process that allows patterns and designs to be made on the surface of metal and glass. An acid resist such as paraffin wax is used to coat the metal surface. The design is then applied by cutting through the wax to expose the metal surface. An acid such as ferric chloride is then used to 'eat away' the material that has been left exposed. Finally, the resistant is removed to leave the original surface texture and a contrast in finishes.

This method is used to make printed circuit boards for use in the electronics industry.

Plating

Plating is a finishing technique which is often used to give metals like brass and copper a coating of a more decorative durable metal such as silver or chromium. The process is carried out by **electrolysis** where the product is charged and the solution acts as a conductor.

Self-finishing

Plastics undergo very little surface finishing because they already tend to be resistant to corrosion and general surface deterioration.

The finish achieved on products such as washing-up bowls, lemonade bottles and plastic drainpipes are all due to the manufacturing processes involved. The high quality of finish is mainly due to the very high quality of the mould.

Texture can be added to the mould which will appear on the final product such as the top of a fizzy drinks bottle. Colour and tone are easily changed with the addition of chemicals and dyes into the plastic material.

The die casting of metal is a similar manufacturing process in that it requires no surface finishing, other than sometimes painting. Molten metal is fed into a very high quality die. Generally, small items such as toy cars and aircraft along with components for domestic appliances are made in this way from lead, zinc, aluminium and brass alloys.

Self-finishing processes, such as **die casting** (see the diagram on page 75), **blow moulding** (see the diagram on page 74), **injection moulding** (see the diagram on page 52) **and extrusion** (see the diagram on page 114), are only suited to high volume production. They are all expensive processes to operate due to the initial cost of the moulds and the machines.

■ **Things to do** ■

1 Collect a range of products and make a note of the various finishes that have been used.

2 What advantages does a robot sprayer have over a manual process when spraying cars on a production line?

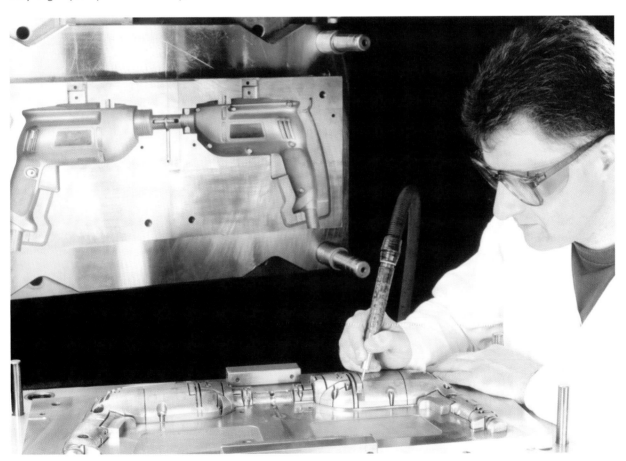

Split mould for injection moulding being finished and inspected.

Marking out materials

Aims

- To understand the need for careful and accurate marking out.
- To understand how the basic tools are used in marking out procedures.

Rules, vernier callipers and micrometers

There are two basic types of rules: steel rules and a steel tape. Both of these start at zero and have millimetre graduations.

When a greater degree of accuracy is needed, vernier callipers or micrometers can be used. The vernier calliper is accurate to within 0.02 mm. The micrometer is accurate up to 0.01mm. They are used for checking dimensions of components during or after manufacture.

Squares

A number of squares are available:

- try square
- engineer square
- mitre square.

Both the try square and engineer square are used for marking lines at 90 degrees to an edge. They can also be used to check that a cut or an edge has been cut at right angles to another. The stock part of the square needs to be held tight against the edge. If light is visible through the join, the edge is not square.

A try square is used widely on timbers, whereas an engineer square is used on metals. Both can be used for marking out plastics.

A mitre square is used for marking out 45-degree angles.

Datum surface/edge

When marking out on wood or metal, measurements should be taken from a datum. A datum edge is a flat face or straight edge from which all measurements are taken. This removes the danger of making cumulative errors.

On a timber surface, the face edge should be carefully chosen. It is then marked with a small symbol for identification purposes. A face side is then selected which is at right angles to the face edge. All measurements are then taken from this side and/or edge.

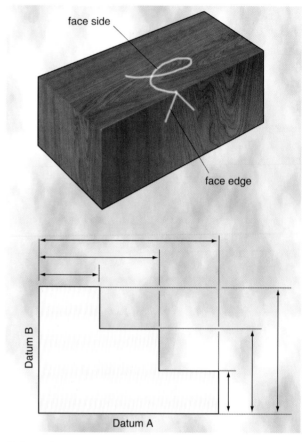

Datum edges

Dividers

Dividers are used to mark circles or arcs on metal and plastic. They can also be used to step off distances. Each of the two legs is ground to a point and because they are hard, they scratch the surface of the metal being marked.

In order to stop the leg slipping on the surface, a small indent is made at the centre of the circles with a dot punch. Acrylic can be marked out in the same way, but for wood you can use a normal compass.

Scribers

A scriber is used to lightly scratch the surface of metal and plastic. It is recommended however that on metal, a coat of **engineer's blue** is applied first.

Punches

Where holes are to be drilled in metal, an indent must be made on the surface with a centre punch to provide a starting point for the drill and to stop it skidding over the surface.

Dot punches are used for marking the centres where dividers are to be used. They are similar to a centre punch, except that the tips are ground to a 60-degree rather than a 90-degree point.

Templates

A template is used when an identical number of shapes or patterns need to be marked out. Any thin material such as plywood or aluminium can be used and drawn around.

Gauges

There are three basic types of marking out gauge:

* marking gauge
* cutting gauge
* mortise gauge.

A marking gauge is used for marking lines parallel to the face edge and side on wood. The gauge consists of a stock that slides up and down the stem allowing various measurements to be set. The gauge should be set using a steel rule that has a zero end. The spur is pushed into the wood as the gauge is pushed or pulled along the length of the timber. It is very important that the stock is held tight against the edge of the timber. This will ensure a parallel line is marked.

A cutting gauge is used for cutting across the grain. It is used in the same way as a marking gauge, but has a blade instead of a spur. The blade cuts the fibres across the grain making it easier and neater to cut with a saw.

A mortise gauge has two pins. One of them is fixed and the other is adjustable. It is used for marking two parallel lines where a mortise and tenon joint is to be cut. The process of marking out is exactly the same as with the two other gauges.

All three of these gauges are used on wood. When marking out parallel lines on metal, odd leg callipers are used. The callipers have a stepped foot which must be kept firmly pressed against the edge of the work being marked out. The hardened point is similar to a scriber that marks the surface of the metal.

setting a marking gauge

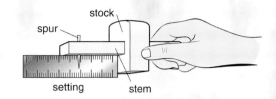

using a marking gauge

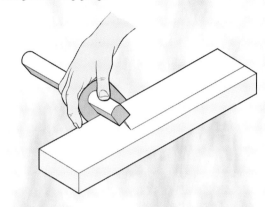

mortise gauge

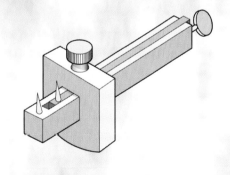

Gauges

■ Things to do ■

1 Explain how a try square is used to check that an edge has been planed square.

2 Why is it important to take all dimensions from a datum edge?

3 What is the difference between a centre punch and a dot punch?

4 Why is 'engineer's blue' used when marking out metal?

Wasting processes 1

Aims
- To understand the term wasting process.
- To understand the various different wasting processes.

A **wasting process** is one that produces waste or unusable material by either cutting bits out or cutting bits off. These processes include:

- sawing
- planing
- filing
- using abrasive papers
- turning wood and metal on a lathe
- drilling
- polishing
- chiselling wood and metal
- milling.

Sawing

All saws are used to cut material that is not needed away from that which is. Every saw makes a gap that is known as the kerf. The kerf has to be wider than the blade so that the blade does not get stuck or jammed. The kerf is made by bending alternate teeth of the blade in opposite directions. When making any cut, you should always cut to the waste side and leave a small amount for finishing. You should also choose the correct saw for the type of material you are using.

Tenon saw

The tenon saw is the most commonly used saw for wood in the school workshop. It is between 250 mm and 350 mm long with 12–14 teeth per inch. All general joints can be cut with a tenon saw. It is specifically used for cutting the shoulder of the tenon in tenon and mortise joints.

Dovetail saw

A dovetail saw is used for small accurate work such as dovetail joints. The dovetail saw is smaller in length than a tenon saw and has 20–25 teeth per inch. This means it has smaller teeth making it ideal for finer work.

Coping saw

A coping saw is used for cutting curves in wood and plastic. It has a thin replaceable blade held in a frame with the teeth pointing backwards. The blade can easily be rotated to cut complex shapes and curves.

Adjustable hacksaw

An adjustable hacksaw also has a replaceable blade that is held in a frame. The blade can be angled to help to cut difficult shapes or if the frame gets in the way of the piece being cut. The blade can vary in length from 250 mm to 300 mm. Blades are available with different size teeth, from 14 teeth per inch for soft materials, or coarse cutting, to 32 teeth per inch for harder materials or finer cuts. The blades are inserted with the teeth facing away from the handle. Whatever is being cut, it is important to have as many teeth in contact with the piece being cut as possible.

Planing

The two types of planes commonly used in a school workshop are:

- smoothing planes
- Jack planes.

Jack planes are used for planing wood flat and to size. They are longer and heavier than smoothing planes, which are used for finishing and planing end grain.

■ Hints and tips ■

- Check the plane is set correctly.
- Plane with the direction of the grain.
- When planing end grain, use a piece of scrap wood to avoid splitting or plane into the centre from both sides.

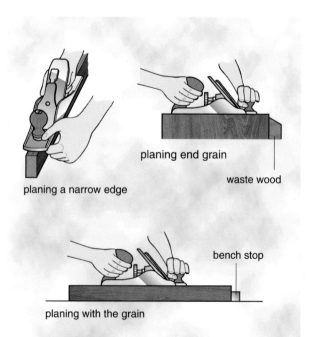

planing a narrow edge

planing end grain

waste wood

bench stop

planing with the grain

Planing

Filing

Files are made from high carbon steel. The main body is hardened and tempered and has rows of teeth. The tang of the file is left soft and fits into the handle.

General work is carried out with a flat file. One of the long edges has teeth and the other edge is left plain and is known as a safe edge. This safe edge will prevent cutting into the face of a square corner. Different profile files are available for a range of applications. There are also other, more specialist forms of file.

There are two basic filing processes used in the workshop:

- cross filing
- draw filing.

Cross filing is used for the rapid removal of waste. The whole length of the file should be used with a downwards force. The file only cuts in the forward direction and it should be lifted off at the end of the stroke and not dragged back across the work piece.

Draw filing is used to remove the marks in the work left as a result of cross filing. This method leaves a much better surface finish because a smoother file should always be used to draw file. An even finer finish can be obtained by wrapping a piece of emery cloth or wet and dry paper around the file and repeating the action.

Chiselling wood

Four basic wood chisels are used in the school workshop. The firmer chisel is a general-purpose chisel which has a square edge. A bevelled edge chisel has a bevelled blade that allows it to get into corners and is especially useful for cutting dovetails. Mortise chisels have much thicker blades and are used with a mallet for cutting mortise joints. Gouges have curved blades and are used for carving.

> ■ **Hints and tips** ■
>
> Chisels should always be kept sharp. Both hands must always be kept behind the cutting edge.

Chiselling metal

Cold chisels are used to cut sheet metal, either by shearing across it or by chopping down on it vertically. They have a hardened and tempered cutting edge while the other end is left soft to absorb the impact from the hammer blows. Different profiles are available allowing access to corners or for producing grooves in the work piece.

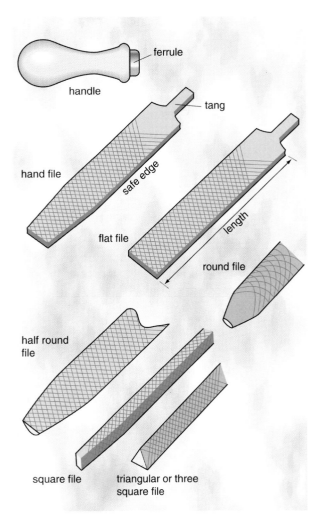

Files

Using abrasives

Abrasive papers are available for woods, metals and plastics. Glass paper that is used on wood is made from glass particles stuck to a paper sheet. The paper is graded according to how much grit there is per unit area. Emery cloth is used on metals and plastics and it is graded in the same way as glass paper. Any form of abrasive paper is best used wrapped around a cork block. This ensures that an even pressure is applied over the work.

> ■ **Things to do** ■
>
> 1 Look at the files in your school workshop. Draw their profile and give an application where they might be used.
>
> 2 Collect a sample of abrasive papers. Stick them in your file. Make notes on their use and try to work out what the numbers on the back mean.

Wasting processes 2

Aim

- To understand machine wasting.

Wood turning

Turning wood on a lathe can be carried out in two very different ways. However, they cannot be carried out at the same time.

The 'outside' spindle is designed to take a faceplate on to which a wooden blank is fitted to make items such as fruit bowls, vacuum forming moulds or fibre-glass moulds.

Turning between centres allows for long pieces of work to be supported at both ends. A fork is used to provide the driving motion to rotate the work piece and the other end is held by a 'dead' centre. Stair spindles and lamp centres are turned in this fashion.

Metal turning

A centre lathe is set up in a similar way to the wood lathe except it has no outside spindle. Work can be turned between centres or held in a chuck which is connected directly to the motor and gearbox.

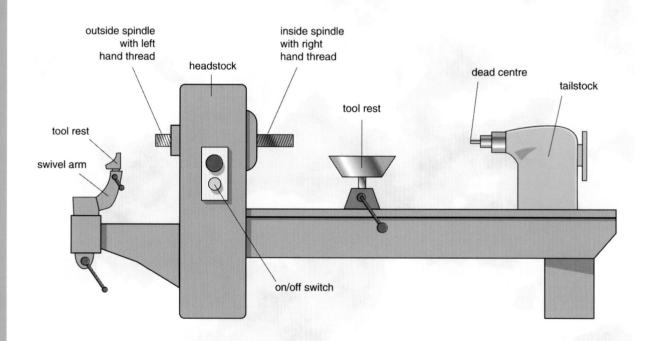

Wood turning on a lathe

Operation		Description
Facing		The tool is moved at right angles to the centre, facing the end surface
Parallel turning		The tool is moved parallel to the centre to form a cylinder
Taper turning		The tool is moved at an angle to the centre to produce a taper
Parting		A narrow tool is fed into the work to trim to length or part the work from the stock bar
Drilling		The tail stock is used as a chuck to hold the drill. As the work rotates the drill is fed into the work
Knurling		A hardened steel wheel is pressed into the rotating work to produce a straight or diamond pattern

Although any shape of work can be turned on the centre lathe, the most common type of work undertaken in school workshops is cylindrical. Therefore the 3-jaw self centring **chuck** is ideal. All three separate jaws move together ensuring the centre line remains constant.

Two basic turning processes are used: facing and parallel turning. Other operations include:

- taper turning
- parting
- drilling
- knurling.

The table above shows the basic operations carried out on a centre lathe.

Drilling

Drilling is a process whereby a rotating drill or bit is powered by hand or an electric motor to produce a hole. Twist drills are the most common tool used for producing holes and can be used in wood, metal and plastic.

Hole saws are used in power drills and are used to cut larger holes from 18 mm to 150 mm in diameter.

Polishing

Polishing can be done either by hand or by using a buffing wheel. A mop on a spindle is coated with polishing compound and then rotated at high speeds. The work is then held firmly against the bottom half of the rotating wheel.

Milling

Milling is a process that can be carried out manually or controlled by a computer to remove waste material quickly and to produce complicated shapes in 2D and 3D. Multi-toothed cutters are used to shape both metals and plastics. Milling can be carried out either vertically or horizontally. In both situations, it is the work that is clamped to a bed that is moved while the cutter simply rotates. Slots, channels and general cutting can all be carried out on a milling machine.

▪ Things to do ▪

1 Make a list of what checks should be carried out before any turning is started on a wood lathe.

2 What is the difference between facing and parallel turning?

3 Collect a number of different drill bits from the workshop and draw and make notes about them.

Deforming processes

Aims

- To understand the term deforming processes.
- To become familiar with deforming processes.

Deforming processes such as laminating, bending and vacuum forming allow the material to change shape without changing its state.

Laminating

Laminating involves building up thin layers around a **former** to produce the desired shape or curve. Thin veneers, or skin ply, are cut to the required shape making sure that the grain is running in the same direction, following the curve.

The thin layers are glued together with an adhesive such as PVA or cascamite. The layers are then trapped between a former or a jig and held under the pressure generated by the cramps. The whole item is left while the adhesive sets. For larger objects, a vacuum table or bag can be used. The layers are prepared in the same way and placed over the former. The air is then sucked from the bag using a pump. The atmospheric pressure forces the layers together and around the former until the adhesive sets.

Bending

Wood, metal and plastic can all be bent quite easily in the school workshop. One easy method of bending wood is to use a technique called kerfing. A series of parallel saw cuts are made along the timber which then makes it easier to bend. The timber can then be fixed to a frame to create a curved surface. MDF is now readily available already kerfed allowing it to be used easily in the home and at school.

Sheet metal can be bent using folding bars. Sheet metal is trapped in between the folding bars which are then held in a bench vice. A rawhide mallet is used to fold the material over.

Tubes can be bent using a pipe bender and a two-part former. The appropriate sized formers need to be set up in the bender and the tube inserted. A former sits over the top of the tube before being tightened down. A large force is then exerted as the pipe is deformed around the former. This process is very useful in the workshop for tubes up to about 18 mm in diameter.

Bends can be made in acrylic using a strip or line bender. An electric current passing through an element or wire applies heat to a very localized area. As the plastic becomes soft, it can be bent into shape or held in a former until it cools.

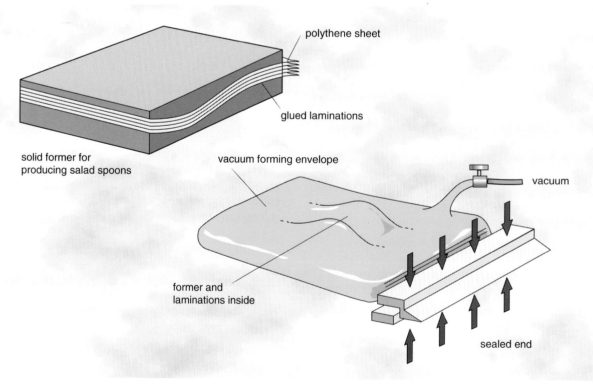

polythene sheet

glued laminations

solid former for producing salad spoons

vacuum forming envelope

former and laminations inside

vacuum

sealed end

Laminating

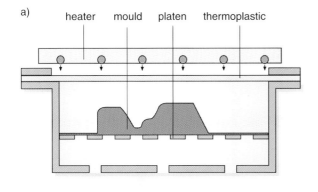

a) heater   mould   platen   thermoplastic

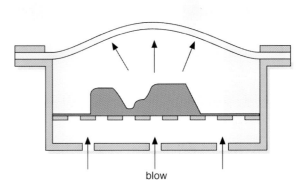

b)

blow

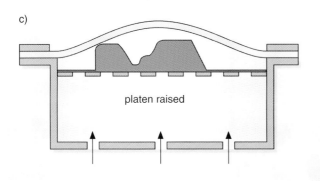

c)

platen raised

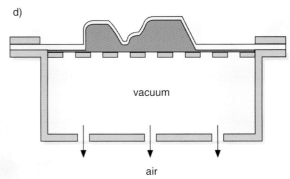

d)

vacuum

air

Vacuum forming

Vacuum forming

The vacuum forming process is used to make various packaging items such as Easter egg containers, packaging for many dairy-based products, trays, dishes and masks.

Generally, the most suitable materials for vacuum forming are thermoplastics such as:

- polythene
- PVC
- ABS
- acrylic.

The process works by reducing the pressure below a softened, flexible thermoplastic sheet. Atmospheric pressure above then forces the sheet down over the mould. The quality of mould design and surface finish will determine the success of the final product. All vertical surfaces need to be slightly tapered and sharp corners should be rounded off. Vent holes should be incorporated to avoid pockets of air becoming trapped.

An example of vacuum formed packaging

■ Things to do ■

1 Look carefully at a vacuum forming mould and consider the detail of vent holes and rounded corners. Try to draw a cross-sectional view for your notes.

2 Look for examples of products which have been made using any of these deforming techniques. Make notes and sketches describing how the techniques have been used.

Fabricating

Aims

- To understand the term fabricating.
- To become familiar with fabricating techniques.

Fabricating is the term used when different pieces or components are joined together to form a single product.

Tapping and threading

Tapping describes the process of cutting an internal screw thread. A hole must first be drilled before the taps are used to cut the screw thread. Three types are used in sequence: taper, second and plug top. The tap is held in a tap wrench and the cutting action involves turning the tap wrench clockwise half a turn and anticlockwise a quarter turn. This action removes the **swarf** build up and prevents the taps from breaking.

Threading is the cutting of an external screw thread. A **split die** is held in a die stock with three screws to locate and to provide adjustments. The tapered side of the die should be used to help start the cut. The same cutting action should be used with half a turn in the clockwise direction followed by half a turn anticlockwise to break off the swarf.

It is essential to align the die and the rod carefully and accurately when you start threading. If the die is not square to the axis of the rod, the thread cut is classed as a **'drunken' thread**.

Wood joints

You should always consider the choice of joint very carefully. It is important to remember that as a natural material, wood will continue to move. Joints can also be used to form features within a product too, at the corners for example.

Dowel joints

Dowel joints are butt joints with dowels used as reinforcement. Dowels are made from beech or ramin. Holes are drilled in both pieces and glue is used to secure the dowels in place and between the joining surface.

Halving joints

Halving joints are made by cutting away half the thickness of the material on each part of the joint. Halving joints can be used on corners, tee, or for cross halvings. Although stronger than a butt joint, it can be strengthened by adding dowels.

Butt joints

Butt joints are the simplest form of joint and the weakest since they only have a small glueing area.

Rebate joints

Rebate joints are also known as lap joints. One end of the two pieces being joined is left plain and the other piece has a rebate cut into it to form the lap.

Housing joints

Housing joints can be cut in natural timber and manufactured boards. They are commonly used in the construction of cabinet work for shelves or dividers. The through housing runs right the way across the panel, whereas the stopped housing stops just inside the front edge. This means that you do not see any of the joint on the front edge.

Mitre joints

Mitre joints are used on picture frames and skirting boards. They allow the mouldings to flow around corners without breaking the clean lines they form. Mitres must be cut accurately at 45° to ensure a good

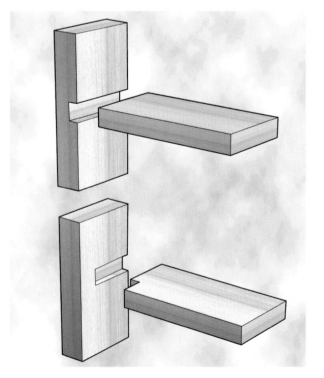

Housing joints

glueing area and a good fit. Once glued, they can be strengthened with the addition of pins.

Mortice and tenon joints

Mortice and tenon joints are widely used in the construction of furniture frames. The mortice is marked out with a mortice gauge and cut with a mortice chisel. The width of the tenon should be one-third the width of the timber. It is a very strong joint.

Rivets

Rivets are most commonly used in sheet metal although they can be used to join acrylic and some woods to metal. Rivets are usually made of the same material as is being joined and are available with a range of heads, the most common being the counter-sunk or round head.

Brazing

All of the heat processes used for joining metal are permanent methods. Brazing is also sometimes known as hard **soldering**.

A gas burning torch provides the heat and the flame is controlled by mixing gas and air on the torch. A **flux**, usually borax, is mixed with water to make a paste and is spread around the joint. The flux prevents excess **oxidation** and helps the **brazing spelter** to flow. Brazing spelter is the filler material which joins the pieces together and is an alloy of copper and zinc which melts at 875°C. Brazing is therefore only suitable for use with mild steel.

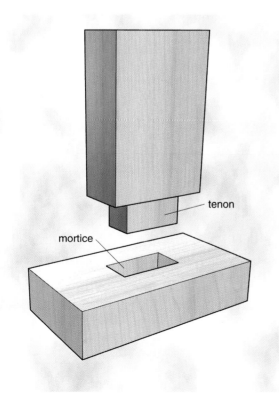

Mortice and tenon joint

Soldering

Soft soldering is a process for making joints in brass, copper and tinplate. When the area to be joined has been cleaned, a flux is used to prevent build up of surface oxides and to aid the flow of solder. A thin layer of solder is applied to each of the pieces being joined. This is known as 'tinning'. When both pieces have been tinned they can be brought together and 'sweated'. This is where a gentle flame is applied and the two timed surfaces join to become one. Alternatively, the joint can be fluxed and heated and then small pieces of solder laid on the joint. This method is better when you are working with difficult and awkward shapes.

Welding

Welding is a process that actually melts and fuses together the two pieces being joined, so the joint is as strong as the original metal.

Various methods of welding are available but probably the most widely used in schools is either electric arc welding or **MIG welding**. Electric arc welding uses a large electric current to jump across a small gap. With a current between 10 and 120 amps enough heat can be generated to melt the metal. A flux-coated filler rod acts as the current carrier and as it is burnt away during the welding process, the flux also burns away and protects the weld from oxidation.

MIG welding is similar to electric arc welding, but it benefits from a continuous feed of filler rod rather than having to replace it. An arc is struck between the work piece and the filler rod and an inert gas flows through the torch to prevent surface oxidation and slag forming.

■ **Things to do** ■

1 Look at a simple piece of furniture like a coffee table or wooden bar stool and try to work out what joints were used in its construction.

2 Make a list of the dangers involved when using any of the joining techniques which involve heat.

Reforming processes

Aims

- To understand the term reforming.
- To become familiar with reforming processes.

A **reforming** process involves a change of state within the material being used. This usually means the material changing from a solid into a liquid or a plasticized state.

Injection moulding

This manufacturing process is a very highly automated process that is used to produce a wide range of common items such as washing-up bowls, buckets and numerous household electrical goods' cases. It is a process which is best suited to **thermoplastics,** but some **thermosetting plastics** can be used.

The **injection moulding** machine is made up of:

- a hopper unit
- a heating element
- a screw and injector unit
- a mould.

The actual process is quite simple and it is repeated continuously to produce moulded products of a high quality that require no further finishing other than removing any **sprue** pins.

- The hopper of plastic granules is used to feed a rotating screw mechanism. As the plastic is moved along the screw thread, it becomes plasticized by the heat from the heating element.
- The rotating screw forces the plastic forward into an area ready to be injected into the mould. The screw mechanism also acts as a ram and this injects the plasticized material into the empty mould.
- The mould is left to cool and sometimes this is achieved by pumping water through parts of the mould.
- When cooled sufficiently, the mould splits open and the moulded product is ejected from the mould by ejector pins.
- The mould then closes and the whole process is repeated over and over again.

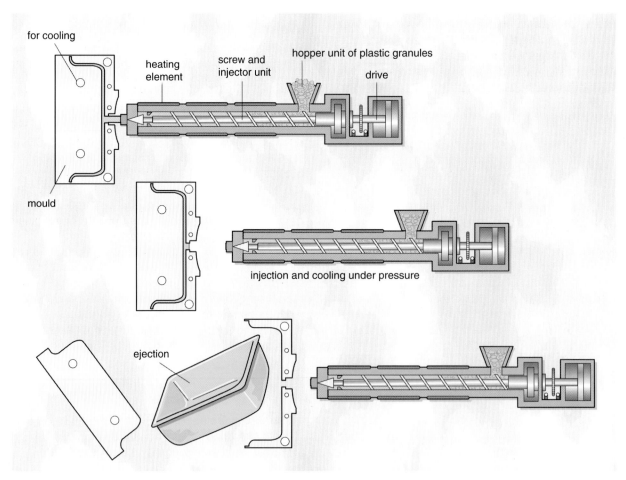

Injection moulding

Casting

Complex shapes like car engine parts and kitchen and bathroom taps are almost impossible to produce by any method other than **casting**.

Casting aluminium in sand moulds is a relatively cheap and simple process which can be carried out in school workshops. The process can be broken down into five main stages:

1 Make a pattern of the required work piece.
2 Encase the pattern in moulding sand.
3 Split the sand box and remove the pattern to leave an empty cavity.
4 Pour molten metal into the mould.
5 Once solidified and cooled, remove the work piece.

Like **vacuum forming**, the quality of final product is dependent upon the quality of the mould/pattern used. Patterns can be either single piece or split. A split pattern must have some dowels inserted to act as location points when lining up the mould. A draft angle should be made on any pattern as this helps when removing it from the sand. External corners should be rounded off and internal corners should have a fillet on them so that sharp corners are avoided.

■ Things to do ■

1 Draw up a list of health and safety issues when casting in the school workshop.

2 What is the difference between a split pattern and a single piece pattern?

3 Draw a flowchart to illustrate the process of injection moulding.

1. The pattern is cut through the centre and fitted with location dowels.

2. One half of the pattern is placed onto a board. The drag is placed over it upside down.

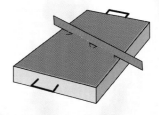

3. The drag is filled with sifted sand and rammed solid.

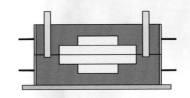

4. The drag should be strickled off with a piece of metal.

5. The drag should be turned over and fitted with the cope. The top half of the mould should be added and the cope should then be rammed up with sand.

6. The cope and drag should be separated, gates cut and the pattern removed.

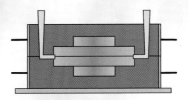

7. Molten metal is poured into the riser until both the runner and riser are full.

8. The final casting with the runner and riser still attached. They can be cut off and re-used.

Casting

Quality of manufacture

Aims
- To understand quality of manufacture.
- To understand the terms tolerance, fixtures, jigs and templates.

Quality

The key concern of any manufacturing industry, whether it is making cars, televisions, food mixers or paper clips, is that of quality. It is not only quality of design that is important but also the quality of appearance and manufacture.

Companies have invested great sums of money in **quality assurance** and in **quality control**.

Quality assurance

Quality assurance involves every aspect of the manufacturing organization in its drive to ensure that the quality of its products exceeds customer requirements.

The ISO 9000

The International Standard of Quality ISO 9000 is awarded to companies that have achieved these high levels of quality assurance.

Quality control

Quality control, however, is related to the actual testing and sampling of products and components during production, assembly and final testing. Quality control is carried out by inspection and testing.

Inspection

Inspection looks at and examines factors such as dimensions, appearance and surface finish. The examination of measurements can be time consuming. However, the use of gauges is an alternative and it is much faster and simpler.

Testing

Testing is more concerned with the function of the product such as:

- Does it work as intended?
- Will it continue to work in two years' time?
- Will it stand up to being dropped?

In some cases, the testing undertaken will result in the product being destroyed. Only a few will be tested in this way due to the expense, but it does ensure that the final product is safe for us all to use.

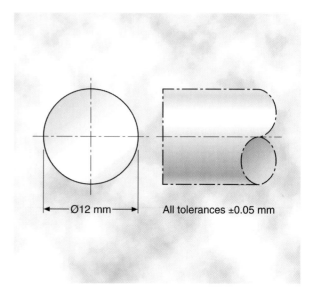

Ø12 mm   All tolerances ±0.05 mm

Tolerances for a shaft

Tolerance

The **tolerance** of a component is the amount of dimensional variation that can be allowed. For some products, such as a cushion for a sofa, the tolerance is not crucial. However, for a rotating shaft in a food mixer that has to fit into a specific bearing, it is much more important.

The shaft illustrated above is 12 mm in diameter with a tolerance of +/−0.05 mm. This means that the shaft could be smaller or bigger by no more than 0.05 mm. If it falls outside this tolerance, then the component would be scrapped. Rather than time consuming measuring with vernier calipers or a micrometer where mistakes can be easily made, a gauge is used.

Jigs and fixtures

In school, you are involved in designing and making one-off products on your own. In industry, where the same product may have to be made many thousands of times, more efficient ways of making things have to be found.

Sometimes this principle will apply to your work, for example if you are having to batch produce a number of identical components such as shelves. This means that you would have to develop a more efficient use of time as well as a method to ensure that the same level of quality was maintained. In this kind of work, accurate and fast repeatability of marking out is required. **Jigs** and **fixtures** are one way of achieving this in industry and in school workshops.

This workshop jig is used for making tee squares

Jigs

A jig is a work holding device that is made specifically to suit a single component. The component would be held firmly in the exact position. The jig is not clamped to the work table, but it is free to be positioned and held by hand.

In the jig pictured above, it is the holding of the two separate components together at a right angle which is important. The two together form a right angle and the small cam locks hold the pieces firm while the hole can be drilled for perfect alignment.

Fixtures

A fixture is similar to a jig, but rather than it being moveable, it is fixed in place so that any tooling lines up immediately without wasting any valuable time.

Fixtures are commonly used on milling machines or drilling machines.

Templates

A **template** is used when a number of identical components need to be marked out. The template is often made from metal and is used by marking or drawing around its edges. Templates are particularly useful when marking out complex and difficult shapes.

■ Things to do ■

1 Design a jig or fixture that could be used to aid the manufacture of the component shown above. Fifty identical components are to be made from softwood.

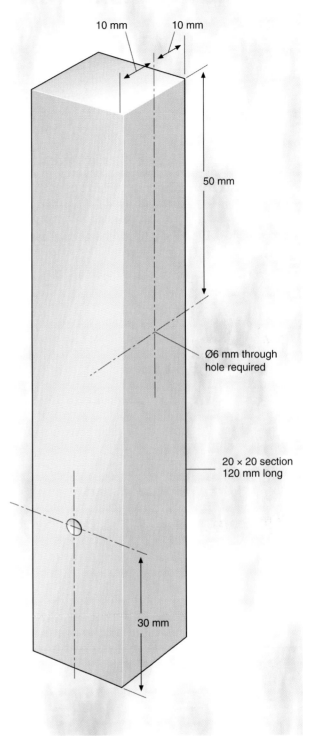

10 mm | 10 mm

50 mm

Ø6 mm through hole required

20 × 20 section 120 mm long

30 mm

Templates like this are useful when marking out complex shapes

Health and safety

Aim

- To understand the need for safe working practices.

In 1974, the Health & Safety Executive (HSE) was formed. This body promotes and ensures that all companies and organizations conform to a certain standard with respect to health and safety. Its aim is to safeguard the health, safety and welfare of employees and to safeguard others, especially the general public. The HSE's work falls into three areas:

- Inspection – it inspects places of work to ensure compliance with legislation.
- Guidance – it publishes documents on a range of subjects for both employees and employers on how to conform to health and safety legislation.
- Updating legislation – it is responsible for keeping up to date and for updating codes of practice.

Health and safety regulations cover a wide range of areas such as temperature, lighting, noise levels, special clothing, disposal of waste materials and chemicals and machine and plant maintenance.

The HSE ensures that the risk workers are exposed to are controlled and minimized whenever possible. Health and safety inspectors have the power to enter factories and premises and to photograph environments and to take samples and recordings as part of an investigation. They are also entitled to see records of inspections and maintenance reports.

The school workshop and environment are also governed by some very strict health and safety legislation. It is very important in your own work at school that you take into account and use safe working practices in a safe working environment.

The most common cause of accidents in workshops is carelessness. The best way for you to create a safe working environment is to follow some basic guidelines on health and safety. Make sure you know:

- what to wear
- how to behave
- how to keep your working environment safe
- how to work safely
- the accident procedure.

What to wear

Loose clothing of any sort is potentially dangerous. All jewellery should be removed before using any machinery. Jackets and blazers should be removed and ties should be tucked in and an apron or workshop coat worn. Long hair should be tied back. Good strong footwear should also be worn and special eye protection should be worn when instructed. Any safety items and equipment used should always be put back for others to use.

How to behave

Quite often accidents in a school workshop occur as a result of foolish behaviour. You should behave sensibly at all times and you should not distract others especially when they are working on machines. You should not rush around the workshop especially when carrying tools and materials.

A safe working environment

It is very important to be aware of potential hazards in the workshop – electrical, heat, chemical or dust. Personal Protective Equipment (PPE) should be used when you are exposed to dust, chemicals or any form of heat treatment. PPE can take the form of gloves, breathing apparatus or masks, ear defenders and eye protection.

Electrical equipment, such as soldering irons, drills and sanders should always be checked before use to make sure they are safe and that they have no frayed, burnt or exposed flex.

When working in the heat treatment area or casting bay, hot tools and work should be left in a cool place before putting them away. You should also let others know that this area is hot.

Many modern adhesives and paints have a chemical base. You should always read the warnings given on the can and take notice of the advice. Masks need to be worn when using GRP, for example, and it must only be used in a well-ventilated area. Tensol cement and contact adhesives should also be used in a well-ventilated area and you should always wash your hands after you have finished working and tidying away.

Safety symbols to be found in the school workshop:

- Mandatory instructions are shown in blue.
- Hazard warnings are shown within a diamond shape.

Safe working practice

It is important to keep your working area clean, tidy and well organized. The area between benches and machines must also be kept clear to avoid potential hazards such as tripping and falling. Tools should be checked before use to make sure that they are in a good safe condition and any that are blunt or broken should be reported to your teacher. Any faults, breakages or damage also need to be reported so that things can be repaired or replaced. When using sharp-edged tools like chisels, you should remember to keep both hands behind the cutting edge to avoid accidents.

Accident procedure

In the event of a fire or an accident, it is important that you know what to do. Your teacher must be informed immediately when any accidents occur. It is important that any injuries, even minor cuts, are properly dealt with. All accidents must also be recorded. In the event of an accident involving someone working on a machine the emergency stop button should be pressed to isolate all power to the machines. A safe and tidy workshop with clear gangways and emergency exits will also help if the building needs to be evacuated quickly.

If accidents are to be avoided, then you need to know how to identify hazards and assess risks. A hazard is anything that might cause harm or damage. The actual possibility of a hazard causing harm or damage is known as the risk. Risk assessment is the study of the hazard and an assessment of how great the risk of that hazard is. Once a risk assessment has been carried out, some measures can be put into place to control the risk. This will ensure that the chance of any harm or damage is less likely to happen.

▪ Things to do ▪

1 Look at the drawing of a school workshop. How many hazards can you spot? What action should be taken to avoid each hazard?

Hazards to avoid in the school workshop

Use of ICT and CAD in single item production 1

Aim

- To understand the use of ICT and CAD equipment in the production of single items.

Clipart libraries

If you are not scanning in and using your own work, clipart illustrations can be used in a document to give it a professional look. These are pre-designed graphic images or pictures that exist on CD-ROM or can be downloaded from the Internet. Clipart can be used as logos, borders or dividers or simply for decoration in your work. Once imported into your work, you can manipulate the image to suit your individual needs. The size can be altered, it can be cropped or framed and the colour can normally be changed.

Scanners

A scanner is very useful for importing images on paper or photographs into graphic files that can then be used and edited on your computer. The most common type of scanner is a flatbed scanner. The quality of the picture is described in terms of resolution, measured in dots per inch (dpi). The higher the resolution, the more dpi, the better the quality of the image. A higher resolution, however, also means a bigger file size, which will take up more disk space.

A scanner should have a resolution of at least 300 dpi and this is a good point at which to start. Scanned images can be taken into most art/graphics-based software packages which means that you can manipulate and enhance the picture.

Digital cameras

Digital cameras may look very similar to ordinary cameras, however, they are able to take photographs without using any film. The picture captured is stored as a graphic file. This file can then be transferred directly on to a computer through a connection lead.

Advantages

A good quality camera is expensive, but you must remember that you will never need to buy another roll of film again. The number of pictures which can be stored in the camera before downloading is dependent upon the size of the disk in the camera.

Digital photography can be used to record any of your work as with normal photography. However, in the same way that scanned images can be imported into graphics packages, digital photographs can also be manipulated in graphics packages. Different parts of the photographic image can be separated, and different colours and textures can be applied.

CAD software

Computer-aided design (**CAD**) covers many different aspects of product design. Quite often CAD is linked with computer-aided manufacture (**CAM**), where components are machined automatically by a computer-controlled machine.

The computer can be used in many different ways from technical research and market research, to the data required to control a large printing or milling machine.

The higher the resolution, the better the quality of the scanned image

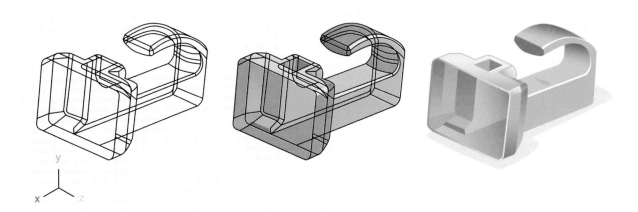

CAD used in the design, development and modelling of a component

Use in industry

In industry, design departments will use CAD and will require data in the form of drawings and details about components and cutting lists. The drawings are entered into the computer by either a mouse, graphics tablet or digitizer, as a series of lines or measurements.

From 2D to 3D

The object being designed is a three-dimensional (3D) object, but it can only be drawn on screen as a two-dimensional (2D) object. From this 2D drawing, a 3D view can often be generated easily.

Standard components

A library of standard components, such as nuts, bolts and washers, will often be available on the computer to the designer. This saves time since they can be drawn once and saved on disk. It is then much easier to load and to import these files into a larger drawing.

Movement

It is possible to rotate the whole object on screen and to view it from a different angle. Computers are now so powerful and fast that they can carry out animations and produce continuous movements of products. They can also generate very accurate coloured renderings of components very quickly. Different colours, surface textures and materials can be quickly applied and the effect seen at once on screen.

Changes

Another benefit of CAD is that changes to the drawing can be made quickly and easily without having to redraw the whole component. This also saves a great deal of time and allows the designers to visualize in a 3D way the effect of small changes in the design.

Printing and plotting

Work and drawings generated in a 3D format can be printed on conventional colour printers such as colour laser or inkjet printers. Work in a 2D format can be plotted out on a flatbed **plotter**. In a standard printer, the paper moves through the machine as the colour cartridge moves backwards and forwards across the paper. On a flatbed plotter, the paper remains still as the pen moves backwards and forwards across the paper. A solenoid is used to lift the pen up and down so that it does not drag as it crosses the paper. A plotter will have a number of coloured pens which are picked up in turn as the computer plots the image. If solid colour is required, then the plotter will draw a series of very close lines together which give the impression of solid colour. It is only possible to use the colour pens that the plotter has. You cannot therefore achieve any graduation of colour to create the effect of a 3D product.

■ Things to do ■

1 Investigate the ICT facilities in your school and look at ways in which you can integrate them in your project work such as clipart libraries, digital photography and CAD.

2 Make a table of advantages and disadvantages of using CAD software.

Use of ICT and CAM in single item production 2

Aim

- To understand the use of ICT and CAM equipment in single item production.

A vinyl cutter

Computer-aided manufacture (**CAM**) is the term used to describe a computer-controlled manufacturing process. CAM is only one part of the whole CAD/CAM system. Once the product or component has been designed, the data, stored as numerical data, is transferred into the CAM system. This link enables the data to be used automatically by the program to machine the component. The profile of the component, created at the design stage, is processed by the software to create a computer numerically controlled (**CNC**) machining program.

The program-generated codes are never actually seen by the designer or the machine setter. Only a few basic decisions need to be taken before the actual machining can begin. These are:

- choice of cutting tool
- depth of cut each pass
- final depth of cut
- cutting speeds.

Once these issues have been decided, the data would be entered into the machine to generate a set of tool parameters for a specific product or component.

Vinyl cutters

A **plotter cutter** is a similar device to a flatbed plotter. The main difference is that with a plotter cutter the paper or vinyl moves backwards and forwards as the pen or cutter travels up and down the length of the machine. As its name suggests, it can also be used to cut as well as to plot. Card, vinyl and other sheet material can be cut using a thin blade, instead of a pen. The pressure of the cut can be adjusted so that different materials can easily be cut. By reducing the pressure of the cut, the blade can be made to score card so that it can be folded for model making and packaging.

Plotter cutters are used widely in the production of advertising signs and banners. Self-adhesive vinyl on a backing paper can be run through the plotter cutter. The fine blade can easily cut the vinyl which can then be removed from its backing paper. This process of waste removal is known as 'weeding' and it is very useful to have a sharp craft knife to help lift the

unwanted material. Depending upon what is considered as waste, either a positive or negative image can be left behind.

It is quite simple to achieve some very professional results on such machines. You can quickly add logos and stripes to any project work as well as developing packaging **prototypes**. Various special materials are available such as 'ceramicon' which is a type of glass material in a very thin film format on paper. When cut out and placed on to a ceramic bowl or plate, it can be fired in a kiln and the image fuses into the surface. Another type of material is also available which can be ironed on to T-shirts, but you must remember to flip your image and to plot it backwards so that when it is printed it is the correct way around.

Milling machines

CNC machines are simply machines that are controlled by numbers. Lathes, milling machines and drilling machines can all be operated as CNC machinery. These types of machines are widely used in industry. Since they can operate in more than one axis, they are capable of cutting straight lines or very complex curves and 3D shapes.

Many different CNC milling machines are available for use in schools, but perhaps the most common type is a dual purpose engraving/milling machine. This can be used to make signs, engrave jewellery and cut wax or foam to make 3D moulds for vacuum forming over. Although this is a relatively small device which sits on a bench top, it will operate in the x, y and z axes.

A machine such as this can easily read in files from different CAD programs. This means that a product or component designed in either Techsoft or ProDesktop can be machined out in either 2D or 3D. The added benefit of a machine like this is that identical components can be cut over and over again without the need for any extra work other than putting the material in place to be cut. The components are also cut very accurately and are left with minimal finishing to be done – such as the polishing of acrylic edges.

Lathes

Lathes can also be controlled by a computer. All the processes which are carried out on a conventional lathe such as facing, parallel and taper turning, drilling and parting can all be carried out on a CNC lathe in plastics as well as metals.

Most industrial machines will have a range of cutting tools set up like the one in the photo with each one carefully positioned to work on a long or batch production run.

A lathe will operate in two axes only with the centre line of the lathe generally being set at zero. Taper turning would see both sets of motors working together in both the x and z axes. As these motors are controlled by a computer, very smooth cuts can be made. This also means that curves such as hemispheres can be cut easily on a CNC, whereas they are virtually impossible on a traditional lathe.

Lathes can be easily set up via a CAD program for a long batch or continuous production run. This means components that have been generated as a result of a profile or half profile being drawn in the CAD package can be easily machined.

A CNC lathe

■ **Things to do** ■

1   Make a table listing the benefits of using CAM for single item production.

A CNC milling machine

1 The drawing below shows a centre punch.

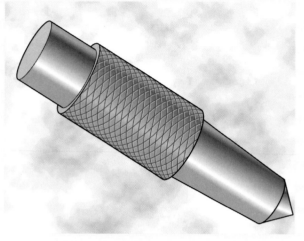

A centre punch

  a Name the two different heat treatment
     processes that the centre punch would
     undergo during its manufacture.
     (2 marks)

  b Explain how the centre punch would be used
     to mark out the centre for the hole to be
     drilled in the piece of metal shown below.
     (2 marks)

2 A screw thread is described as being M10 for
   example. The thread illustrated below is a
   'drunken' thread.

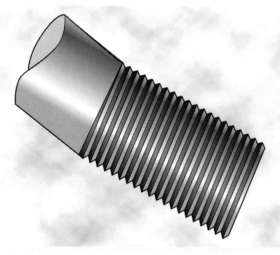

A drunken thread

  a List two reasons why the thread is drunken.
     (2 marks)

  b Explain using notes and sketches how to cut
     an internal thread in a piece of metal 20 mm
     thick. **(4 marks)**

A piece of marked out metal

Section C
Manufacturing commercial products

Product manufacture

Aim

- To understand levels of production including one-off, batch and high-volume.

One of the major factors to consider when designing any product is the eventual retail selling price and the level of production needed to make the manufacture economically viable. It is cheaper to make larger quantities, so it may be necessary to adapt the product's design to suit the demands of **mass-production** manufacturing techniques.

One-off production

At the other extreme of the production scale, there is **one-off production**. These are items that are produced singly as a one-off. There may be a number of reasons why a product falls into this category of production. It may be a stadium for the Olympics or the World Cup, a bridge to span a particular point on a river, a communications satellite, etc.

In each of the above examples, safety and reliability are more important than the manufacturing costs. Each individual project allows the designers and engineers freedom to create a stunning design such as the Blinking Eye bridge over the Tyne in Newcastle.

Although it is a footbridge, it is in very stark contrast to the Tyne Road Bridge pictured behind it in the photograph below.

There are many other one-off items such as jewellery or personal clothing designs such as a wedding dress or a tailored suit. In each of these examples, **aesthetics** (the look of the item) is more important than the cost and function.

Other examples of one-off design could be a shop, room or office. These are spaces to be fitted rather than products and the design of lighting, colour, wall covering, display and flooring all need to be considered and the designer needs to take into account what the space is to be used for, and any practical considerations as well as cost.

Custom-made furniture is another example of one-off production.

Client needs

When designing a one-off product, the relationship between the client and the designer is very important. Usually, the client will have some very clear ideas about his or her own needs and what the product must do. The designer will be aware of the limitations and characteristics of the materials.

The Blinking Eye footbridge and the Tyne Road Bridge in Newcastle

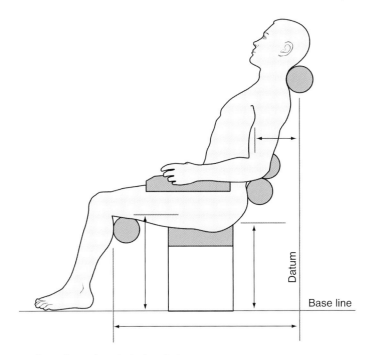

Key dimensions when designing chairs

Concept ideas

It is the responsibility of the designer to come up with some initial concept ideas. He or she will present these to the client for discussion and either further suggestions or approval. In terms of the designer's creative ability, the design of a one-off product is often a great opportunity to be able to demonstrate his or her flair to the full. The designer is in a position to be able to challenge ideas and preconceptions about what a product must look like such as:

• Why does a chair have to have four vertical legs?
• Why does it have to have a back support?

Choice of materials

The choice and use of material is also very important when designing and making a piece of one-off furniture. The combination of marble and glass, smooth cool materials, with dark grain patterns makes for a pleasing contrast. Stains, varnishes and paints can all be used to enhance and customize a piece of furniture. There is also the difference between gloss and matt finishes to consider.

In some cases, natural forms of materials can be used. Sometimes a piece of timber is identified by the designer and he or she is able to see its potential at once and to design with the specific material in mind.

Ergonomics and anthropometrics

If the product is a piece of custom-made furniture, for example, the size can be tailored to meet the individual and specific dimensions of the intended user. The relationship between size and the individual is called **ergonomics**. The science concerned with the measurements and the measuring of people is known as **anthropometrics**. In ergonomics, anthropometric data is used to ensure that products will be safe and comfortable to use.

If a chair was being designed for the majority of people, the designer would base his or her design on standard recommended dimensions. In order to design to the precise size of an individual, the designer would have to test and record the specific dimensions and make a **mock-up** of the person to represent them.

■ Things to do ■

1 Carry out a survey of your class to collect some anthropometric data such as height and reach. Try to work out what the average is.

2 Make a list of the key dimensions you would need when designing a one-off chair for an individual.

One-off manufacture

Aim

● To understand the nature of a one-off piece of design.

Case study: Michelle Watling, wedding accessories designer

Wedding dresses and accessories are almost all created as one-off pieces of design. This case study looks at the work of Michelle Watling, a designer and creator of some unique head-dresses and wedding-day accessories.

In this example, the client requested something different and unusual for her wedding-day head-dress. In the initial stages, the design focused around royal crowns with much use being made of books showing royal crowns from history.

Although there was discussion and consultation between the designer and the client, this was difficult because they were several hundred miles apart. Drawings, pictures and letters were exchanged and eventually the client was satisfied.

Small sample sections of the head-dress were made and sent to the client for comment and approval. Catalogues were used to select buttons and beads, and colour cards and the wedding dress fabric were used in order to match the colours.

It was difficult to draw up detailed drawings of the items used and the construction techniques involved. The final product was painstakingly made by hand from twisted silver wire of differing gauges. In the early stages, the main ring was tested for size on the bride's head and small loops were added to the base circle, so that the crown could be fixed to her head using hair grips.

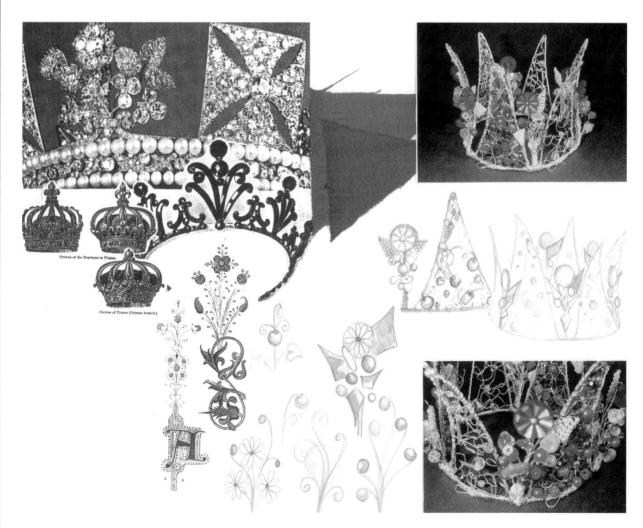

Initial thoughts and designs for wedding head-dress

Table light based on the Eiffel Tower

In total, 17 hours were spent on the actual realization and the bride was very pleased with the final result. This product would simply be too difficult to make in any quantity. It is very much a one-off.

Using jigs, formers and templates

Much of your GCSE project work involves one-off manufacture, although needs and situations are often identified that could allow for products to be produced in greater quantities. Even though one-off manufacture is undertaken, it is still necessary at times to use **jigs**, formers and **templates** to aid in the manufacture.

For the light pictured above, jigs were used to ensure that all four curved legs were identical. Another jig was used for drilling holes in the legs and making sure that the small cross-members lined up. A final jig was used to hold the legs in the correct position while they were being soldered together. A template was made to help to cut the fabric to make the shade inside.

Case study: bedroom storage furniture

The pale pink bedroom storage cabinet pictured left was designed specifically to go into a bedroom. The shape and symmetrical curves give the product a very individual look. The choice of colour fits in with the general colour scheme of the bedroom. The sides are hollow and a template was used to make sure the curves were identical. Thin, skin ply was stapled across the width to give the impression of a solid shape. The strength of the whole piece was in the rigid construction of the two vertical shelf supports. Although not flat-pack, a few changes in the basic design would have allowed for the product to be collapsed for easy transportation and storage.

Bedroom storage furniture

▪ Things to do ▪

1 Think about your GCSE project and about how jigs, formers and templates might aid the manufacture.

Batch production

Aim

- To understand the nature of design in relation to manufacturing in batches.

'Batches' of identical products

Batch production means an industrial manufacturing system which produces a fixed quantity or 'batches' of identical products. It is not normally size specific and a batch may be small or large depending on the economics of the product being produced.

Specific tools

Generally, tools and machines will be set up specifically to produce the parts and components needed for the product. Once the batch has been produced, the machines and tools will be used to produce parts for other products. The specific tools and any specially made tools will be stored away for use at a later date should another batch of identical components or products be required.

Quick response

One advantage of batch production is that it often allows a very quick response to customers' demands since all the tooling is in place. The workers who operate the machines are generally very skilled and efficient in setting up machines. An example of where it is beneficial to be able to respond very quickly to a situation is in the food industry. This industry often faces demands or peak periods during different types of weather, such as increased demand for ice cream in hot weather, and the producers need to respond very quickly.

Economics

Batch production fits in between **one-off** and **mass production** despite the fact that a batch may run into several thousand products or components. The economics behind a batch size is quite complicated, but it can be explained in relation to manufacturing costs.

Break-even point

There comes a point below which it is not viable to produce a certain quantity based on the set-up costs, which take into account tooling, and this is known as the 'break-even point'. Below this it is too expensive, above this the cost of manufacturing decreases which makes the product more affordable in the retail market.

Market viability

Obviously, careful market research prior to making a decision about a product's market viability will give some indication of the number that could be sold.

Cash flow

Batch production allows companies to have greater control over their cash flow. They do not have to invest lots of money into the manufacture of products which will then be stored for long periods in warehouses or retail stores. If very careful attention is paid to stock levels and stock control, not too much money has to be invested into stock. Many companies now make extensive use of ICT to control stock levels and to undertake automatic ordering of new stock.

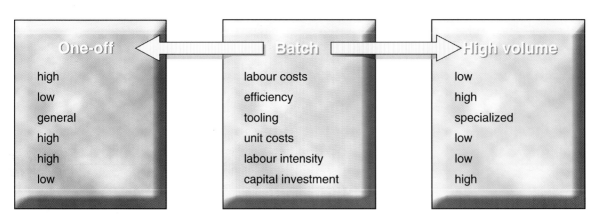

One-off	Batch	High volume
high	labour costs	low
low	efficiency	high
general	tooling	specialized
high	unit costs	low
high	labour intensity	low
low	capital investment	high

How costs change as the level of production increases

Virgin offers a wide range of goods and services and is easily recognized by the unified use of a distinctive typography, colour, packaging and staff uniforms. This allows many items to be batch produced as new stores are opened or services are expanded.

Manufacturing costs

Not all machines and manufacturing processes are suited to batch production. The more manual the process, the greater the manufacturing costs because of the labour costs. If, however, products such as telephone cases are required, they can be injection moulded despite the initial high investment cost of the mould. They can be batch produced when required and the **injection moulding** machine can quite easily be set up to run through a batch.

Large companies

A large multi-national company requires many designed and manufactured items. Many of these items are batch produced as new stores are opened or existing stores are refitted to present a new image. Staff uniforms, trolleys, advertising, shelving and display systems, checkouts, service counters and store interiors must all be designed and manufactured.

Common features

There are a number of common features and strands, such as colour and font style, which will be consistent throughout the whole company. This consistency contributes to a corporate image that makes the brand recognisable regardless of where in the country or world you may be shopping.

From the moment you enter the store, signs, icons and display frames conform to the corporate image. The frames will be batch produced regardless of the graphic image they hold. The same applies to the trains and aircraft, for which the tables and chairs have been produced in styles and numbers that make batch production viable.

Many of the items in a high street store will have been designed to allow components to be produced economically in batches. For example, production systems are simple and effective. Machines will be set up quickly and the numbers required will make efficient use of machine time.

Corporate identity

Corporate image is very important to this type of company. Recognising a brand makes the customer feel familiar and comfortable with it. The hope is that, as a result, customers will then feel loyalty to that brand, whether they are on the high street, or in other parts of the country or even in other places around the world.

This approach is adopted by most chain stores on the high street. The design of fittings and layouts is something that is consistent throughout all stores, helping to promote a brand identity.

■ Things to do ■

1 Why is corporate image important for a company like Virgin?

2 Look around your school and at its promotional material and letterheads. Try to identify whether it has a clear corporate image and make a presentation of what you find.

3 Draw up a table of the advantages and disadvantages of batch production.

Examples of batch production

Aim

- To understand how batch design and manufacture can lead to a range of products.

Case study: MFI flat-pack furniture

Another example of **batch production** is the manufacturing of flat-pack furniture.

MFI, a high street chain of stores which sells household furniture, sells and stocks a wide range of flat-pack furniture for bedrooms, living rooms, offices and kitchens. Particular environments such as kitchens and bedrooms offer a wide variety of choice in terms of style, finish and pattern. However, the main body (the carcass) is the same across most ranges. Kitchen units for example come in a standard range of sizes. The extract below from a store magazine shows the range of sizes and the prices across the range.

A standard 1000 mm wall unit has the same dimensions and position of holes regardless of the range it belongs to. It is assembled in an identical fashion so the instructions are relevant to the whole range too. The position and depth of the slot for the back panel and the position and size of the holes for shelf supports and hinges are identical. The style of the doors may be specific to the range, but again, the dimensions and positions for the hinges are identical.

This element of design makes it ideal for batch production. The cutting of a number of identical pieces, the machining of grooves and the drilling of holes make batch production the ideal process.

The only aspect that is different is the actual material from which the product is machined. It is this which makes the range seem comprehensive.

Since all MFI furniture is distributed on a national level rather than held in local stores, stock control and levels can be carefully monitored and demands responded to quickly. The opportunity for new ranges, colours and finishes is, therefore, due to the nature of the product and the manufacturing processes and techniques involved.

Case study: hanging basket brackets

Jigs and **fixtures** have a very important role to play in batch production. This case study is based around the design and batch production of a number of identical hanging basket brackets which were to be sold at a school fair.

Flat strip Bright Drawn Mild Steel (BDMS) was used since it is a cheap material which can be quite easily worked and joined in the school workshop environment. Two different sections were chosen. The larger

		500 F/Ht Open Wall Unit, Framed	290 F/Ht Open End Wall Unit, Modern	290 F/Ht Open End Wall Unit, Gallery	290 F/Ht Open End Wall Unit, Spindle	310 F/Ht Open End Wall Unit, Framed	500 Full Height Plate Rack	500 Full Height Framed Plate Rack	500 Larder Tall Unit	600 Sin... Tow...
		W5FOP *1	WEND	WEND	WEND	WEND	W5PLA *1	W5PLA *1	T5LAR	T60V
		H720xW500xD290	H720xW290xD290	H720xW290xD290	H720xW290xD290	H720xW310xD290	H720xW500xD290	H720xW500xD290	H2170xW600xD575	H2170xW...
		Unit price	Unit price	Unit price	Unit price	Unit price	Unit price	Unit price	Unit price	Unit price
MXC1 *VALUE PLUS* MEXICO		–	£39	–	–	–	–	–	£64	£
HTG1 *VALUE PLUS* HASTINGS		–	–	£38	–	–	–	–	£85	£
SRL1 *VALUE PLUS* STIRLING		–	£27	–	–	–	–	–	£50	£
OKD1 *VALUE PLUS* OAKLAND		–	£38	–	–	–	–	–	£50	£
HFD1 *VALUE PLUS* HATFIELD		–	–	–	£58	–	£56	–	£67	£
CWY1 *VALUE PLUS* CONWAY		£58	–	–	–	£34	–	£64	£88	£
WSM1 *VALUE PLUS* WESTMINSTER		–	–	£38	–	–	–	–	£79	£
RIM4 *VALUE PLUS* RIMINI		–	£38	–	–	–	–	–	£79	£

		Unit price	SALE	Unit price	SALE	Unit price	SALE	Unit price	SALE	Unit price	SALE	Unit price	SALE	Unit price	SALE	Unit price	SALE	Unit price	SALE	Unit price
WKI1 NEW WOODSTOCK	50%OFF	–	–	£82	£41	–	–	–	–	–	–	–	–	–	–	£180	£90	£188
CDG1 CADOGAN	50%OFF	–	–	–	–	£82	£41	–	–	–	–	–	–	–	–	£180	£90	£188
WSS1 WESSEX PINE	35%OFF	–	–	–	–	£59	£38	–	–	–	–	–	–	–	–	£185	£120	£187
CTW2 CHATSWORTH/RVL2 RAVELLO	45%OFF	£152	£83	–	–	–	–	–	–	£91	£50	–	–	£164	£90	£219	£120	£221
MTR1 NEW METRO	50%OFF	–	–	£66	£33	–	–	–	–	–	–	–	–	–	–	£240	£120	£242
HDN1 HUDSON	50%OFF	–	–	£72	£36	–	–	–	–	–	–	–	–	–	–	£254	£127	£88
SOM2 SOMERTON	40%OFF	–	–	–	–	£64	£38	–	–	–	–	£178	£106	–	–	£219	£131	£221
MDN2 MADISON	50%OFF	–	–	£82	£41	–	–	–	–	–	–	–	–	–	–	£262	£131	£264
MHT1 MONTREAL	50%OFF	–	–	£72	£38	–	–	–	–	–	–	–	–	£170	£101	£262	£131	£264
ADL3 ARUNDEL	40%OFF	£151	£90	–	–	–	–	–	–	£94	£56	–	–	–	–	£234	£140	£241
STK1 NEW STOCKHOLM	50%OFF	–	–	£76	£38	–	–	–	–	–	–	–	–	£170	£102	£262	£131	£264
OKY3 OAKLEY	40%OFF	£147	£88	–	–	–	–	–	–	£94	£56	£139	£83	–	–	£234	£140	£241
ASO1 ASCOT	40%OFF	–	–	–	–	–	–	£69	£41	–	–	–	–	–	–	£134	£80	£146
HNL1 HENLEY	45%OFF	–	–	–	–	£70	£38	–	–	–	–	–	–	–	–	£146	£79	£159
ACN1 ANCONA	40%OFF	–	–	£82	£41	–	–	–	–	–	–	–	–	–	–	£151	£90	£159
CDZ1 CADIZ	45%OFF	–	–	£75	£41	–	–	–	–	–	–	–	–	–	–	£164	£90	£172
NBK1 NEBRASKA	45%OFF	–	–	£75	£41	–	–	–	–	–	–	–	–	–	–	£164	£90	£172
NPT1 NEW NEWPORT	50%OFF	–	–	£82	£41	–	–	–	–	–	–	–	–	–	–	£180	£90	£188

MFI sells flat-pack furniure in a wide range of 'standard' sizes

Batch producing scrolls

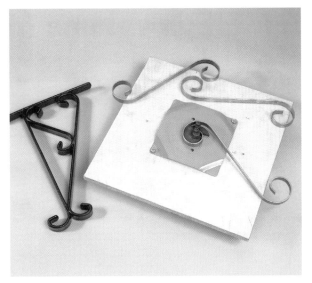

Different sized scrolls are produced for use on the final product – a hanging basket bracket

was used for the vertical strips and the smaller for the decorative scroll work.

Both section sizes were available in 1800 mm lengths, so the brackets and individual pieces were designed to make the most economical use of the material. The two sections were also capable of being cut to size on a small guillotine. The larger section could also have holes punched through it rather than drilling.

A small selection of scrolling formers were available for use and a variety of different sized scrolls could be formed. Because mild steel is quite malleable, it can be easily **deformed** cold using the formers and pressure exerted by hand. A small mark on the former indicates the point at which to stop bending.

Once the vertical wall bar and the two separate scrolls were formed they were joined. **MIG** or arc **welding** is an ideal process for this and it is a quick, safe and efficient method to use in the school workshop. A small jig was designed and made to ensure that each individual piece was held in the correct position. In addition, the tolerance involved was not that critical.

Once the bracket had been welded in the jig on the one side, it was flipped over and welded on the other side. The brackets were then cleaned and degreased before being either dip coated or painted with Hammerite.

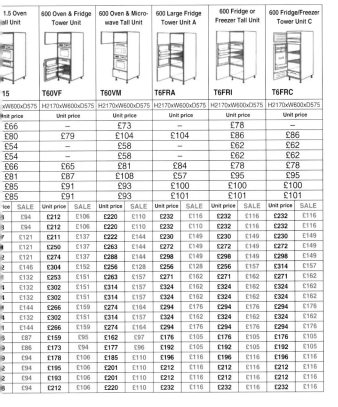

1.5 Oven Tall Unit	600 Oven & Fridge Tower Unit	600 Oven & Micro-wave Tall Unit	600 Large Fridge Tower Unit A	600 Fridge or Freezer Tall Unit	600 Fridge/Freezer Tower Unit C
15	T60VF	T60VM	T6FRA	T6FRI	T6FRC
xW600xD575	H2170xW600xD575	H2170xW600xD575	H2170xW600xD575	H2170xW600xD575	H2170xW600xD575
Unit price	Unit price	Unit price	Unit price	Unit price	Unit price
£66	–	£73	–	£78	–
£80	£79	£104	£104	£86	£86
£54	–	£58	–	£62	£62
£54	–	£58	–	£62	£62
£66	£65	£81	£84	£78	£78
£81	£87	£108	£57	£95	£95
£85	£91	£93	£100	£100	£100
£85	£91	£93	£101	£101	£101

ice	SALE	Unit price	SALE	Unit price	SALE	Unit price	SALE	Unit price	SALE	Unit price	SALE
8	£94	£212	£106	£220	£110	£232	£116	£232	£116	£232	£116
3	£94	£212	£106	£220	£110	£232	£110	£232	£116	£232	£116
7	£121	£211	£137	£222	£144	£230	£149	£230	£149	£230	£149
4	£121	£250	£137	£263	£144	£272	£149	£272	£149	£272	£149
2	£121	£274	£137	£288	£144	£298	£149	£298	£149	£298	£149
2	£146	£304	£152	£256	£128	£256	£128	£256	£157	£314	£157
1	£132	£253	£151	£263	£157	£271	£162	£271	£162	£271	£162
1	£132	£302	£151	£314	£157	£324	£162	£324	£162	£324	£162
1	£132	£302	£151	£314	£157	£324	£162	£324	£162	£324	£162
1	£144	£266	£159	£274	£164	£294	£176	£294	£176	£294	£176
1	£132	£302	£151	£314	£157	£324	£162	£324	£162	£324	£162
1	£144	£266	£159	£274	£164	£294	£176	£294	£176	£294	£176
6	£87	£159	£95	£162	£97	£176	£105	£176	£105	£176	£105
9	£86	£173	£94	£177	£96	£192	£105	£192	£105	£192	£105
9	£94	£178	£106	£185	£110	£196	£116	£196	£116	£196	£116
2	£94	£195	£106	£201	£110	£212	£116	£212	£116	£212	£116
2	£94	£193	£106	£201	£110	£212	£116	£212	£116	£212	£116
8	£94	£212	£106	£220	£110	£232	£116	£232	£116	£232	£116

■ Things to do ■

1 Describe why jigs are useful when it comes to assembling identical products in batches.

2 Produce a **Gantt chart** for the production of these hanging basket brackets.

High-volume (mass) production

Aim

- To understand the nature of design and manufacture in relation to high-volume production.

High-volume production is another name for **mass production**. Both of these terms describe the process where the same product is produced continuously. Some products produced in this way, such as fizzy drinks bottles or plastic washing-up bowls, require little or no further assembly. Other products, such as a television remote control or a mobile phone, require a number of separate pieces to be assembled by hand along an assembly line before they are ready for use.

This type of assembly process is often described as a production line. In certain circumstances where this assembly process extends over a 24-hour day, it is called continuous or flow production. Quite often, car manufacturers operate this type of assembly where their workers work shifts allowing for 24 hours a day assembly.

When products or components are produced in high volumes, they are often produced on very expensive machinery and require very expensive tools and moulds. The high initial 'tooling up' costs have to be recovered in the overall cost of the product.

Certain manufacturing processes lend themselves to high volume production. Although the tooling costs and the actual machinery are very expensive, generally, the actual manufacturing costs and material costs are a lot lower.

Case study: plastic model kit

The plastic model kit pictured was **injection moulded** using a very complicated and expensive mould. The cost of the raw material can be measured in pennies. The actual retail price must reflect:

- the initial investment in the machinery
- the making of the moulds
- the raw materials
- packaging
- transportation and distribution
- profit.

It is likely that one company was responsible for the entire manufacture of the model kit although it would probably have had the instructions printed elsewhere along with the packaging.

Not all companies make all the components of their product though. Mobile phone manufacturers and major motor companies do not tend to make each individual part. Car manufacturers, for example, often buy in a large number of the components needed. They might buy tyres from a specialist tyre manufacturer. The car manufacturers would, however, be responsible for the assembly. They would also have to

Mass-produced plastic model kit

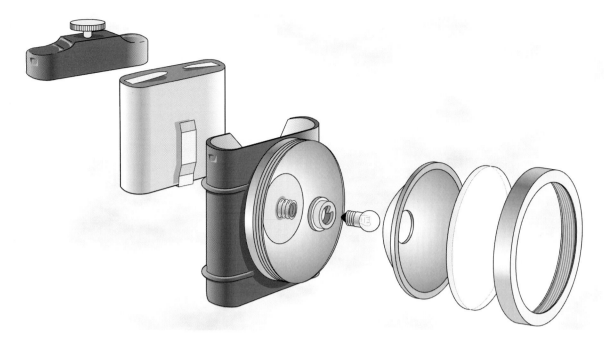

Exploded view of a cycle light

ensure that each of their individual suppliers could guarantee the quality of their work and supplies in line with ISO 9000, the **quality assurance** system.

Case study: bicycle lights

Bicycle lights have been restyled in recent years and have become much slimmer and more aerodynamic. As a product, they are made up from several separate components as the exploded diagram above shows:

- a lid with a battery contact
- a battery
- a main body
- a bulb
- a reflector lens
- a glass cover
- a screw-on cover cap.

With the exception of the glass covers, the bulb holder and the battery, all the parts will have been **injection moulded**. This is a highly automated process which is ideally suited to high-volume production runs. Although the initial tooling up of the moulds is expensive, the long-term manufacturing costs are low.

The bulb holder and the battery contacts will have been made from brass, a very good electrical conductor. Brass tends to tarnish (reacts to the air around it) less than copper and it is also slightly tougher.

These various electrical components will be assembled manually on a production line. Once the whole product has been assembled, it will be wrapped in a protective plastic film and packed in a box.

Modern cycle lamps have a different shape and use a different type of light. Most use super bright light emitting diodes (LEDs). They provide a much more intense source of light and are also more efficient in terms of the electrical current they draw from the battery. Some lights also contain simple electrical circuits which make the LEDs flash on and off, making them more obvious to drivers. This type of light tends to be the rear light which is coloured red.

Lenses are also now made from plastic rather than glass. This allows them to be injection moulded too. A surface texture is moulded into the lens creating a kind of prism effect. As the light hits the surface it is shone into the prism, creating a more intense light source.

These features have all been developed in an attempt to make the cyclist more visible and therefore safer. The improvements have been made possible by developments in materials technology and manufacturing processes.

■ Things to do ■

1 Dissemble a cycle lamp and look at the different pieces from which it is made.

2 Explain the different ways in which the parts are joined together.

3 Examine the different materials which have been used and give a reason why they were chosen. Look especially carefully at the lens and the pattern inside it.

Examples of mass-produced items

Aim

- To understand how manufacturing processes are used to create high-volume production.

Blow moulding

Blow moulding is a process that generally produces hollow products such as plastic ducks for the bath and water butts. Blow moulding of plastic products is a very highly automated process. Like all of the high-volume production processes it is extremely expensive to set up and to produce the moulds.

With a product such as a sit-on toddler's toy, identical bases are produced in high volumes. Different textile covers are made to fit the one base so that there seems to be a wide product range. This means that the same toy can be made to appeal to a wide market sector, such as boys and girls and different age groups.

This is a very good application of a standardized design, the base being used to create a large product range. In economic terms, it also extends the life of the mould since it will be used in more than one product.

How blow moulding works

- The mould is made up from two separate halves which are normally symmetrical. This type of mould is known as a split mould.
- A thin plastic tube known as a parison is formed and fed down into the open split mould.
- The mould then closes around the parison.
- An injector blows air under pressure into the parison forcing the plastic to the sides of the mould.
- This also helps to cool the plastic and, therefore, to speed up the overall process.

Casting

The process of casting is not always associated with high-volume production. However, **die casting** is a type of casting that does lend itself to high-production levels. Unlike sand casting, where the cope and drag need to be separated after each product has been cast, die casting uses a metal mould similar to those used for blow and injection moulding.

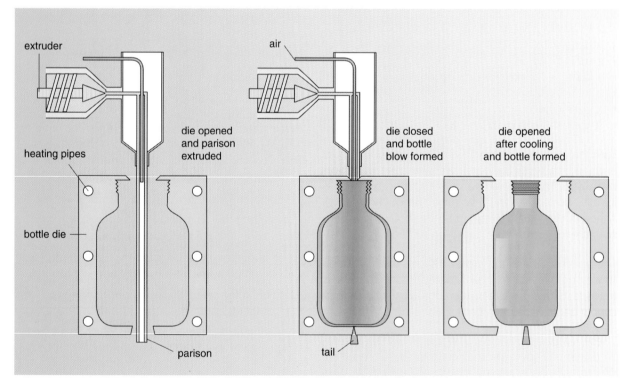

extruder

air

heating pipes

die opened and parison extruded

die closed and bottle blow formed

die opened after cooling and bottle formed

bottle die

parison

tail

Stages of blow moulding

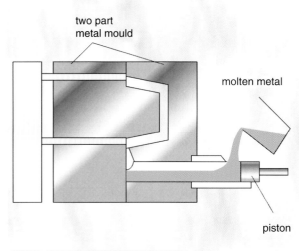

two part metal mould

molten metal

piston

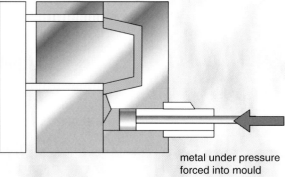

metal under pressure forced into mould

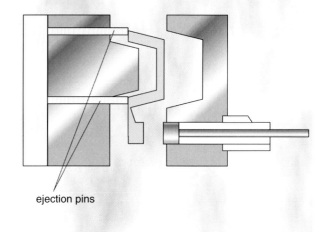

ejection pins

Pressure die casting

Die casting uses low temperature **alloys** usually based on zinc. There are two different methods of die casting:

- gravity die casting
- pressure die casting.

In gravity die casting, the molten metal is poured into the mould and gravity helps the metal to flow around the mould. In pressure die casting, the molten metal is forced into the die under pressure. This method is used where there are quite complicated products and components being cast.

The last few pages have looked at various levels of production and given examples of products that fall into each of the three.

Much of deciding whether something should be batch, mass produced or made as a one-off is to do with economics. Balancing the method of production with its scale is an important decision to make.

In your own project work, one-offs are usually considered although some products are worthy of batching or mass production. Where this is the case, a one-off is made as a prototype.

▪ Hints and tips ▪

It is important that you consider the industrial applications that your project may be subjected to. You should record these details in your project folder.

▪ Things to do ▪

1 Look at the list of manufacturing processes below, and say whether each one is most suited to one-off, batch or high-volume production:

a blow moulding

b CNC milling

c carving wood

d forming metal scrolls

e cutting mortise and tenons by hand.

2 Would the following products be produced as a one-off or would they be batch or mass produced? Explain each of your answers.

a DVD player

b medals for the Olympics in 2008

c a national sports arena

d a display stand for a supermarket chain

e a kettle.

Use of ICT and CAD/CAM in batch and volume production

Aim

- To understand how computer systems enable fast communication.

Communication systems

Communication systems have become very advanced and continue to improve all the time. Networks now exist in almost every environment where there are computers and with a modem you can connect your computer to the Internet, send e-mails and send and receive faxes.

What's a modem?

A modem takes digital data from one computer and translates it into sound (or analogue) before sending it down a telephone line into another modem. The modem at the other end converts it back into digital data.

E-mail

One of the most useful features of the Internet is e-mail. E-mails are transmitted at staggering speeds. A message can be sent to reach another computer on the other side of the world in just a few minutes. The only cost that is involved is the price of a short local phone call.

Almost all e-mail programs allow you to write messages offline (that is, without actually being connected to the Internet). Working offline will save money while you are composing your message. Then when you send the message, the computer will connect to the Internet (go online) and your message will be sent.

In some e-mail programs, you can create an address book. This saves time because you can simply click on the name you want to send a message to rather than typing in the e-mail address every time.

It is also possible to send the same e-mail to a number of recipients. This is similar to posting mail to a large number of people, only it is electronic mail. A supplier of nuts and bolts might, for example, have a special offer or want to clear some old stock. Rather than sending out letters or telephoning around his customers, he could e-mail them all with details.

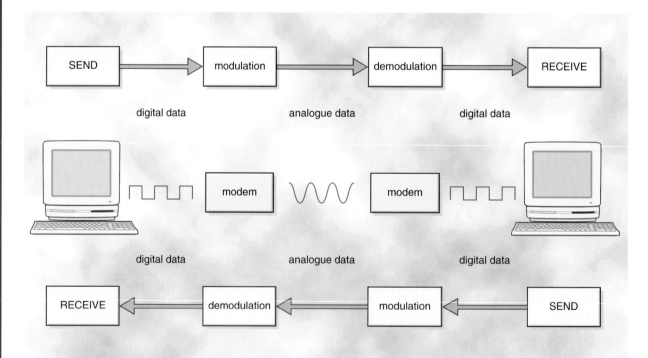

*How a modem works – 'modem' comes from the words **mod**ulator and **dem**odulator. A modem turns computer data into a type of signal that can be transmitted down a telephone line. A similar modem is also needed at the receiving end, carrying out the reverse process. Most computers have a built-in modem*

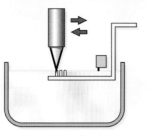

1 Start: laser draws first layer on to resin

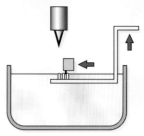

2 Wiper moves across work to create an even surface

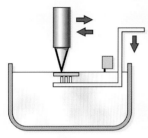

3 Platform moves down with solidified first layer. Laser draws second layer on to resin. Wiper moves across work to even surface

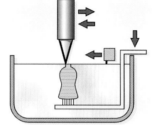

4 Process continues until all layers are produced

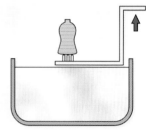

5 End: completed part is withdrawn from resin vat

The process of stereolithography

Attachments

It is also possible to send pictures and other types of computer files such as databases, text files and spreadsheets as attachments to an e-mail message. The recipient receives the attachment as a separate file which he or she simply clicks on to open.

It is possible for companies to send files in the form of attachments to their clients. Designs, prices and production schedules can all be created in various software packages and saved as separate files. They can then be e-mailed as attachments to the clients anywhere in the world. Immediate responses can be put together and replied to at once, saving time and money.

Sending designs and rapid prototypes by e-mail

Some specialist companies are now capable of producing rapid **prototypes** in 3D using a special process known as **stereolithography**. This is a complicated process in which a liquid resin is hardened by a laser beam passing across a tank. As the resin is tapped off, a hardened product is left.

It is possible to design a product such as a new kettle shape on a CAD package and to e-mail the design to this type of company. The company will then produce a rapid prototype model that can be sent back to the design team to look at, touch and feel. This use of e-mail has enormous benefits for both of the companies involved. Large files can be sent quickly and dealt with in very short spaces of time.

■ Things to do ■

1 Explain two advantages for the designer of using attachments when communicating via e-mail.

2 What benefits are gained by the client and the designer when using technology such as stereolithography?

The Internet

Aims

- To understand how the Internet can be used for gathering information.
- To understand web browsers and how to use them.

The world wide web (www) is the public face of the Internet. It is a huge collection of websites made up from individual web pages. The graphic images and text you see on screen make up part of a web page.

Web pages are usually written in a language called HTML (hyper text mark-up language). This allows pages to be linked together and for a collection of pages to be joined together to form a website. One website can be linked to other sites around the world to form a global world wide web of connected sites.

In order to be able to find your way around all of these websites or to 'surf the net', you need a web browser. There are two main browsers:

- Microsoft Internet Explorer
- Netscape Navigator.

How to find information

Once you are connected to the Internet, the quickest way to find information is to type the web address of a relevant site into your browser address box and to press the 'return' key. The first page that appears is the 'home page'.

If you do not know the specific address, you can use a search engine to look for information. A search engine will search for key words or categories that you select. You will then be presented with a list of sites to visit. There are numerous search engines available, but at *www.searchenginewatch.com* they explain how the main engines work and how efficient they are.

Some search results will suggest further sites that are linked and may be relevant. These are shown as underlined text. By clicking on the underlined text, that link will be activated and a new website and name page will be presented. You do need to be careful since you can end up getting sidetracked from your original search.

The Internet is an almost unlimited resource bank with huge amounts of information and data accessible to all. It is a very useful tool to use when working on projects. It is also useful for businesses to keep up to

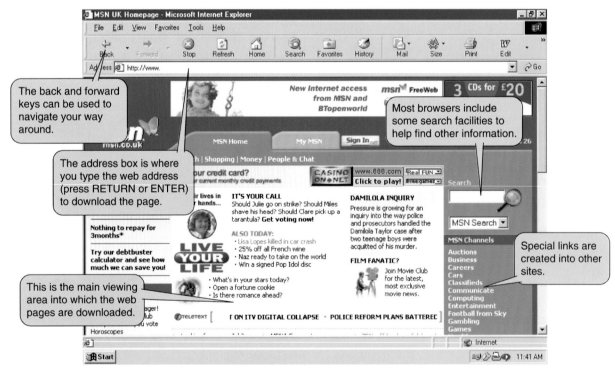

Web browser

date on what their competitors are doing. When putting together your own project work you could use the Internet to research:

- existing products
- materials
- suppliers
- manufacturing companies
- contact with experts.

Most websites have e-mail links built into them so that you can automatically contact them. This might allow you to request a brochure or further details about a product range. It is also an opportunity to be able to make contact with individuals in companies who might be able or willing to help you with information.

Remember

When e-mailing companies, be polite, use correct terminology and compose (write) your e-mail like a letter. Be professional with its presentation and ensure your spellings are correct.

▪ Things to do ▪

1 Use the following search engines to conduct a search for one of the topics listed below:

www.altavista.com

www.excite.co.uk

www.google.com

www.yahoo.co.uk

  a   plastics

  b   injection moulding

  c   MDF.

2 a   What differences do you get in the results when you use different search engines?

     b   What does this tell you?

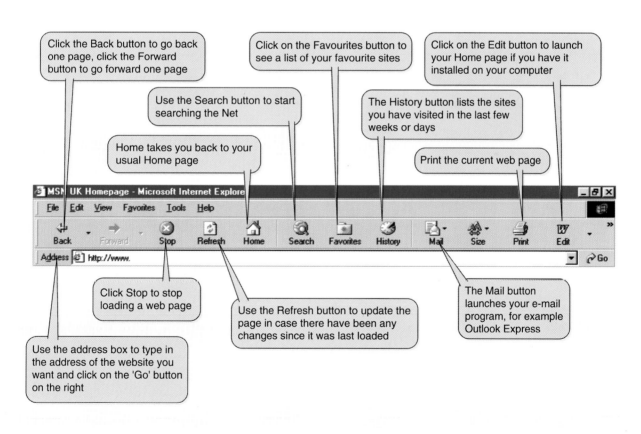

Tool bar on the web browser

EPOS systems

Aim

- To understand what EPOS means and how it is used by manufacturers, suppliers and retailers.

Electronic point of sale (**EPOS**) systems are used to gather and record information about products. Each product has its own unique black bar code that is either stuck or printed on to a label and attached to the product.

country of origin	manufacturer/ company	specific item	check

The bar code is made up of a series of black lines on a white background. The strips vary in their width and in the spaces between them. A number is also printed below the bar code in case the strips get damaged and cannot be read by the laser scanner. The code comprises a 13-digit number which gives information about the country of origin, the manufacturer and the specific product code.

The EPOS system has been described as the 'intelligent' till in supermarkets. As the product is passed over the laser scanner, the bar coded item is identified from its unique code and a series of events is triggered. The information is stored directly on to a computer at the same time as all the other products are being scanned at the other checkouts. The computer records what has been sold and monitors sales across the whole store. It will also have a record of how much of a certain item is currently in stock. When a certain point is reached it will re-order electronically by e-mail from the distribution centre.

Companies can use the information gathered by EPOS systems to analyse sales and stock levels. Over the period of a year they can look at how any product has sold at any particular time. The system records and works out how much money the shop has made, giving the company more control over theft, wastage or damage. It also generates an itemized bill and till receipt.

Manufacturing industries are now adopting the same sort of stock control principles as used by the EPOS system in their own production and stock control systems. During the manufacturing process, products are given a unique bar code just like a product in a supermarket. Indeed, some companies have developed their own coding systems for specific components that are used in the assembly of a product. Fixings such as screws or pop-rivets are boxed into a

An infra-red beam from a scanner reads the bars and spaces of a barcode by measuring the reflected light

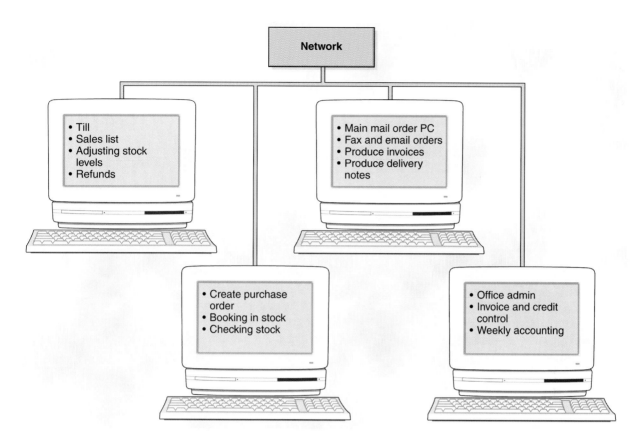

An EPOS system set up for use in a small business

batch of 50, for example, and are then bar coded. This process is repeated for all the individual components needed for the entire product manufacture.

Boxes are then booked out for assembly and can be tracked throughout the whole assembly process along with all the other components. Replacements are automatically ordered since stock levels are recorded and can be acted upon at once.

This recording and tracking process also allows companies to monitor the assembly of their products much more carefully with respect to quality control and quality assurance. They also benefit by not having to hold massive stocks of bought-in components which ties up money.

Bar codes are easily produced for internal use within a manufacturing environment. EPOS keyboards are used to create the code and a dedicated EPOS printer is used to produce the labels or stickers. They can then easily be fixed to stock boxes or products when completed.

Scanners are available in a number of forms with the two most common types being the hand-held and counter-mounted.

The diagram above gives an indication of how a business can be set up to operate its own EPOS system. Although this system uses four separate computers, a very small business can operate a whole system on just one computer.

The various terminals would be set up in different parts of the factory such as the sales department, stock room or warehouse, and the office or accounts department. With all machines networked, the whole system is kept up to date.

The EPOS system has many benefits for supermarkets and manufacturing industries, including:

- quick and efficient sales and ordering processing
- search by bar code facility to check stock levels
- adjust and record stock levels on a daily basis
- daily reports generated, for example, on sales history and stock valuation
- holds customers' and suppliers' details.

■ Things to do ■

1 Look at a range of products from one particular manufacturer and see if you can identify similar parts of the bar code. What might they be used to signify? For example: Dulux – paint, varnish, filler, wallpaper paste, brush cleaner.

2 What do the initials EPOS stand for?

3 What advantages does an EPOS system bring for a supermarket?

Aim

- To understand how CAD/CAM systems can speed up manufacturing and make manufacturing processes more flexible.

Computer integrated manufacture (CIM)

In recent years, manufacturing companies have had to become ever more efficient and competitive. As a result, companies have been forced to review and revise their working practice and procedures in order to develop ideas and to get them into the market place quicker.

The development of design is ultimately about creating better and updated products. This can be because of advances in technology, because a product can be made cheaper or because production methods and technology have improved.

The advance of computers has been of enormous benefit to designers and manufacturers. Designers using **CAD** systems can create, develop, record and communicate with others anywhere in the world. **CAM** systems are able to translate design data into codes and programs into manufacturing data. This enables **CNC** machinery to cut or turn products and components automatically, quickly and accurately.

Computer integrated manufacture (**CIM**) has become a much more effective way of integrating CAD/CAM systems. CIM links together the two separate areas of CAD and CAM.

Drawings are automatically downloaded to CNC machines. Robots control the movement of material from one machine to another and remove them when complete. Automatic chucks on lathes and milling machines are controlled by computers. Hydraulics or compressed air provide the force to hold the work tight while being machined.

The making is carried out within a **manufacturing cell**. Very little human intervention occurs other than for servicing. Often in larger companies, AGV (automatic guided vehicles) are used to carry completed work away into another cell for either storage or assembly. For more on AGVs, see page 86.

CIM also includes quality control, stock control, parts handling and distribution. The whole manufacturing process can be operated 24 hours a day, 7 days a week without the need for lunch and tea breaks. It makes for a very efficient way of manufacturing.

Managing product and design data

With much design now being carried out on computer-based systems, data can all be stored electronically. This data can be used to control a series of machines and robots in the manufacturing processes. It can also be used for stock control and for production planning in drawing up work schedules.

When drawing up work schedules, the computer is able to take all factors into consideration. While being able to access data about materials and machining times, the system is able to calculate times and identify where any clashes may arise. Attention would be drawn to this point and production staff would be able to

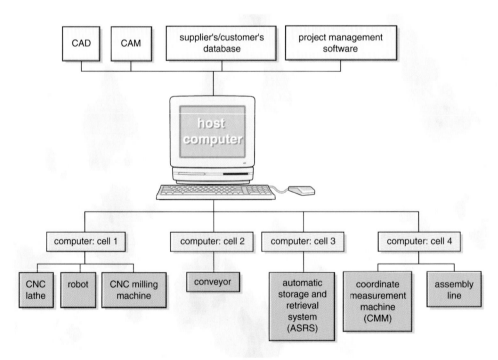

A CIM layout

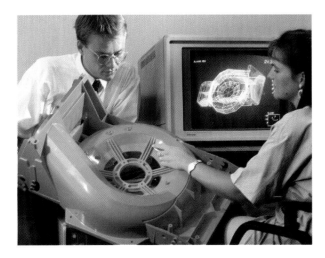

Engineers used CAD to design a plastic component for a lawnmower. CAD allows the component to be viewed from any angle and small changes can be made rapidly. The effect on the rest of the design is calculated almost immediately. This plasic component has been produced using injection moulding

resolve the problems before going into production. The other advantage is that with data about orders on file, the computer is able to juggle work schedules based on order priority. The subsequent sequencing of manufacturing and delivery of resources is set by the overall priority.

As more departments within an organization have greater access to the design and product data, it is the responsibility of each of those individuals to ensure that their input conforms to the overall quality assurance procedure. With effective computer communication systems, a company can move towards a total quality management (TQM) system. This is a quality procedure to which all employees have a commitment and they must ensure that the computer system and data management conforms to this standard too.

Design data in the form of materials' technical data would also be stored and accessed by engineers and machine setters. They would ensure that the correct speed and feed settings are programmed into automatic machines.

Managing stock control

Trends in manufacturing have seen products manufactured more quickly and in the exact quantities to meet customer demand. Traditional manufacturing would have measured productivity in relation to the raw materials being converted into finished products. It is now common practice to measure productivity against the level of demand: money in against money out. This has led to a type of manufacturing and stock control known as 'just-in-time' (JIT).

JIT has improved efficiency, but it is essential that all components and materials arrive on time since any delay will result in increased costs. JIT requires good relationships with suppliers, very careful planning and accurate time schedules for each and every stage of manufacture.

Before JIT, companies needed to stockpile components and materials. Although this meant that production was never held up, it meant that money was tied up in expensive stock. It also took up space around the factory which added to the expense. Traditional stock rooms have now been replaced by JIT manufacturing and stock control is managed by computers and comprehensive planning schedules. Good stock control now takes the form of having the right materials, in the right quantity in the right place at the right time.

■ Things to do ■

1 Describe the advantages to a company of using the JIT approach to manufacturing.

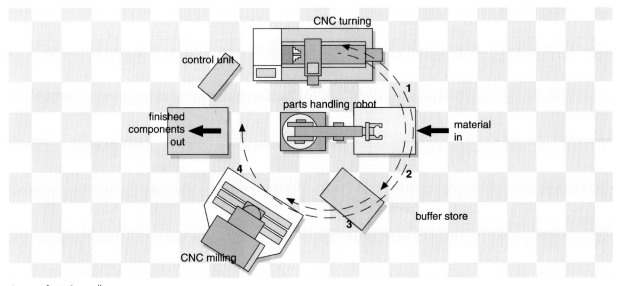

A manufacturing cell

Aim

- To understand how 2D and 3D images are used to create virtual products.

2D modelling and the creation of 3D 'virtual' products

Computer graphics and the use of computers in terms of CAD can be broken down into four different areas:

- Modelling – how objects are described and represented in terms of lines, arcs, solids and colours.
- Storage – how images are stored in the computer's memory.
- Manipulation – how models can be changed and altered in shape and size or the bringing together of separate drawings.
- Viewing – the model can be seen from a particular view point.

In most software packages, the design and actual drawing construction are undertaken in a 2D format. Some comprehensive packages like ProDeskTop are very effective at turning a 2D drawing into a 3D product. The initial transformation is first carried out using a wire frame to represent the 3D product.

Because the product has not yet been colour rendered, it remains quite a small file. At this stage, the image can be very quickly moved, rotated or even animated.

Once any final changes and amendments have been made, the wire model can be fully colour rendered creating a solid model and 'virtual' product on screen. At this stage, the product can be given any number of finishes and can even be seen in a range of materials such as brass, cast iron, steel or acrylic.

This virtual modelling has allowed designers and engineers to look at 'real' products and components even before they have been made.

The software packages are quite simple to use once the basics have been mastered. They allow you the flexibility to be able to draw and to model in 2D as well as having the ability to generate a 3D view. An added feature is that they can also be linked into a variety of CAM machinery such as small benchtop engraving and milling machines.

Fast, accurate and repeatable production processes

One major benefit of any CAD/CAM system is its ability to be able to repeat images once drawn. Depending on the software, images can also be revolved, scaled up and down, mirror imaged, squashed or stretched in order to find the best shape.

The general ability to be able to copy and then duplicate images can save an enormous amount of time for the designer. The component illustrated in the picture on page 85 is part of a table leg assembly. For each table, four identical components are required.

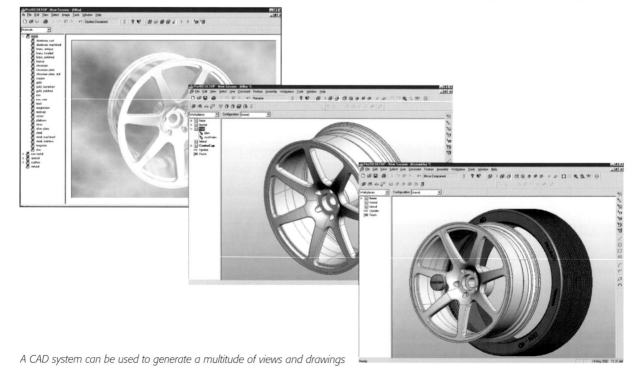

A CAD system can be used to generate a multitude of views and drawings

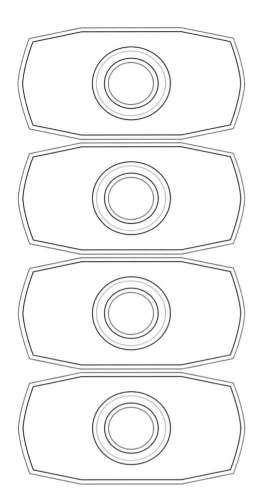

The repeated component has been 'nested' to ensure that material wastage is kept to a minimum.

Once one component had been drawn, it was copied and then duplicated to form one single drawing with four identical parts ready for machining. They have been placed as close together as possible to reduce the waste material. This process is known as 'nesting'.

Just as a CAD system can be used to copy and duplicate drawings, CAM can be used to cut multiple copies of any single component.

CAM allows for the fast and accurate cutting out of identical components once the machine has been set up with the correct feed speeds and tool offsets. Given that only the new material needs to be put in place and the machined components removed, the CAM set-up can operate for 24 hours a day, 7 days a week.

Although CAD/CAM systems are expensive to purchase and install, they are very cost effective in very long, high volume production runs.

Advantages of CNC machinery

- Reduced set-up costs of machines – CNC machines do not need jigs and fixtures.
- Removes need for expensive form tools.
- Human error and scrap reduced significantly.
- Manufacturing costs more predictable.

Disadvantages of CNC machinery

- High initial set-up costs.
- Expensive training of programmers and operator.

▪ Things to do ▪

1 Explain the advantages of creating 'virtual' products on screen.

2 Describe how the level of accuracy has been improved by using CAD/CAM to manufacture the table leg components pictured opposite.

Production control

Aim

- To understand how production control uses CNC machinery and equipment for automatic production.

Essentially, two different extremes of manufacturing exist. At one end, there is a dedicated automatic machine which is capable of making hundreds of thousands of an identical product such as a fizzy drinks can. This is a very cheap method and produces products of a consistently high quality.

Initially, this seems a good investment. However, the machine must be kept running in order to make it financially viable. Also this highly automated machine may appear to be inflexible as every component is identical.

At the other extreme, a skilled operator and craftsman/woman makes whatever is required whenever required. This method is completely flexible, but the main problem here is the rate of production and the consistency. As a result of the time taken production costs are very high.

The arrival of **CAD/CAM** saw the introduction of the flexible manufacturing system (**FMS**). This was both flexible and automatic. FMS was achieved by combining **CNC** machines such as lathes and milling machines with robots for the handling of materials. Cutting tools such as the lathes and milling machines are controlled by very powerful computers. Their movements and other functions such as coolant pumps, spindle speed control and automatic work holding devices are controlled by instructions or data in the form of a program. AGVs (automatic guided vehicles) operate the transportation of materials and finished products within the system.

Most machines and machining is simply carried out following a set routine or program. This type of system is known as an 'open' loop system. A system that can respond to the position of a piece of metal to be picked out of a bin and place it in a vice relies on the feedback of information of its own position. This type of system is known as a 'closed' loop system. It is a highly automated system and relies upon powerful computers to control it. When a machine such as this is used in a FMS, it can operate independently and with other fully automated machines such as lathes, quality control devices and material storage facilities. An FMS is illustrated below showing how the various parts make up a comprehensive flexible manufacturing system.

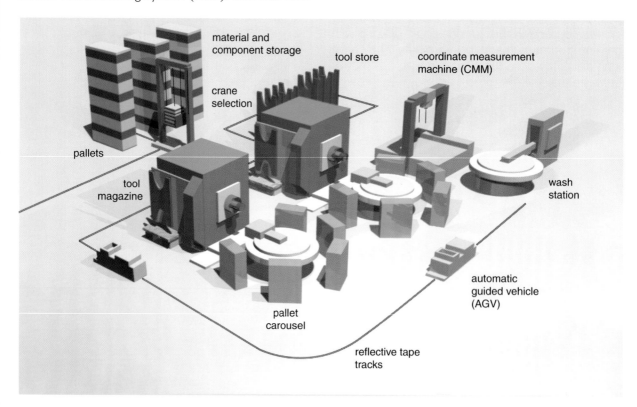

A flexible manufacturing system

Quality control

Quality control (QC) can best be described as 'techniques for checking quality against a standard or to a tolerance'. It involves the inspection of products and seeks to spot products that are not up to standard.

The inspection process generally takes place after an event or process within the manufacturing process. Quality control inspections are often carried out by carefully trained inspectors within the company or organization.

Dimensional accuracy is a key area for a quality control inspector. A dimensional **tolerance** is the allowed or acceptable variation in the dimensions of the part. This may be height, weight, diameter, depth or angle. With all manufacturing, there is a limit to the dimensional accuracy. The nearer the limit is reached, the more expensive it becomes to manufacture. A compromise has to be reached which balances cost against dimensional limits.

Variations in sizes can be caused by:

- tools wearing
- variations in materials
- changes in the environment
- labour variations – different workers.

Checking of work can take place at four different periods during the whole manufacturing process:

- checking incoming materials
- checking of purchased parts and components before they are added into the assembly line
- checking of work in progress – any defaults can be reworked or rejected
- finished products – finding a fault at this stage can result in the whole product being scrapped.

Almost all checking is carried out on a sampling basis by taking out at random one part in 50, for example.

A technician inspects the engine of the Air Force Thunderbird's F-16 Falcon by using CAD. With CAD, the engine's systems can be checked faster and more accurately

Checks would also be carried out on identical parts made by different workers or on different machines. A table or chart would be used to record the data and to make judgements about what is inside or outside any tolerance limits. For example, a component may have a tolerance range of +/–0.05 mm. Any records made above or below the upper and lower limit are outside the range and not acceptable. The components which are inside the two limits are acceptable and will move on to the next stage.

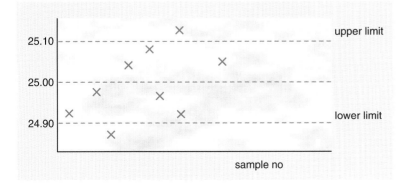

A scatter graph can be used to show upper and lower limits

■ Things to do ■

1. Why is it important to check identical components made by different workers?

2. What possible environmental fluctuations could cause changes in the dimensions of a material?

3. What can be done to components which are above the upper limit on a scatter graph?

Practice examination questions

1 a List three advantages of using e-mail to a manufacturing company.   **(3 marks)**

   b Explain how the Internet can be used to gather information.   **(2 marks)**

2 a What do the initials EPOS stand for?   **(1 mark)**

   b List three benefits for manufacturing industries which use EPOS systems.   **(3 marks)**

3 a Explain the meaning of the term quality control.   **(2 marks)**

   b Variations in sizes of components made of identical machines are caused by a number of reasons. List four reasons why such components may differ in size.   **(4 marks)**

4 The child's toy shown at the bottom of the page is made by a process known as pressure die casting.

   a Using notes and sketches, explain the process of pressure die casting.   **(4 marks)**

   b Explain the main difference between the following processes:
   • pressure die casting
   • gravity die casting.   **(4 marks)**

A miniature car made by pressure die casting

Section D:
Design and market influences

Product safety

Aims

- To understand the importance of product reliability and safety.
- To understand the various safety standards.

Concern for safety

Any product can be regarded as an object or system that has been designed and produced in response to a human need. Cars, televisions, mobile phones, furniture and CD players are all examples of product design.

One car may be quite similar to another in a range. It might be quite different to another made by a different manufacturer. What is common to all cars, and to all products, is a concern for the safety of the user and the reliability of the product. Product safety and reliability are very important issues for both the consumer and the manufacturer.

Consumer confidence

When we buy a new product, we usually take it for granted that it will work and that it will be safe. This is because before a product gets to this stage, the designers and manufacturers have worked hard to ensure that they design and make products which work safely and efficiently.

Extensive testing

Manufacturers undertake extensive tests to ensure and guarantee that their products are safe to use and reliable in operation. When James Dyson was designing and developing his revolutionary method of vacuuming carpets, the product was tested exhaustively. Not only was the 'cyclonic' action tested for reliability, but the switches, revolving brushes and the electrical system were also tested. For example, basic tests would be carried out on the on-off switch by repeatedly switching it on and off. This tests the reliability of the switch and gives an indication of how long it can be expected to last.

Fair comparisons

Any testing must be fair if comparisons are to be made with similar existing products. Most tests that look at a range of products will also give some indication about performance against value for money. The Consumer Council publishes reports of its tests and results of almost every new product that enters the market place. The results are published in the magazine *Which?*

Safety testing of a new car

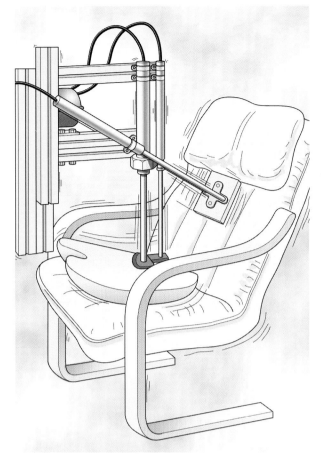

Test rig

Special testing equipment

For certain products, special pieces of testing equipment also have to be designed and made. The home furnishings retailer IKEA prides itself on the design and quality of its products. In its stores, there are test rigs on display demonstrating the testing of products.

Analysis and evaluation

In the design process, there is always a stage at which existing products are analysed and evaluated. The designer attempts to improve the look, performance or safety of the product.

Once a prototype has been designed and made, it will be tested for safety by the manufacturer. Consideration will be given to the different types of people who might use it and in what type of conditions and environment it might be used.

British standards

When the manufacturer is happy that its product is safe, independent safety inspectors then carry out a series of tests on the product. Specific products and types of product also have to conform to and pass very stringent British Standards (BS). There are sets of British Standards for the testing of a whole range of items. Each British Standard has a reference number and date of publishing or amendment.

BSEN 50088: 1996 Incorporating Amendment No1

This British Standards document deals with the safety of electric toys. It covers:

* protection of cords and wires
* screws and connections
* resistance to heat and fire
* making and instructions.

In order to show that a product has met the requirements of the specific standard/specification, certain marks may be affixed to the product or packaging.

The BSI Kitemark

The BSI Kitemark shows that the product has been tested by BSI Product Services and meets the requirements of the relevant standard/specification and, in addition, that the manufacturer's quality system meets the requirement of ISO 9000. Only once a product has met all the requirements is the Kitemark issued. The products continue to be tested a minimum of twice per year and the manufacturer is visited, usually twice per year, to ensure continuing compliance with ISO 9000.

The CE marking shows that the product meets the requirements of the relevant EU Directive(s).

The CE marking

▪ Things to do ▪

1 List the potential safety hazards of an electrical toy.

2 Imagine you are the manufacturer of a kettle. List what tests you would carry out on it to ensure it is safe and reliable.

New technology

Aim

- To understand the use of new and smart materials.

Carbon fibre

New materials and new uses for new materials are continually developing. The technological advances and materials technology in the Formula 1 racing industry have been leading the development of lighter, stronger materials for many years now.

Carbon fibre is a composite material. A fibrous material, known as the reinforcement, is suspended in a plastic resin called the matrix. Carbon fibres are used in very much the same way as glass reinforced plastic (GRP). They are, however, much stronger.

The carbon fibres exist in several forms, as:

- a loosely woven fabric
- a string of filaments wound together
- a non-woven mat of short fibres
- loose short strands.

Carbon fibre products and components are very strong and have a very high strength to weight ratio. This means that they are very strong in comparison to their weight. Carbon fibres are used in golf clubs, tennis rackets, skis and bikes. Carbon fibre composite materials can also be found in the rotor blades of helicopters.

Kevlar

Kevlar is a composite material based on carbon fibres or made with the fibres of the special polymer Kevlar.

Kevlar has many structural uses in the components of aircraft, leading to major reductions in weight. Kevlar components are as strong as aluminium, but have weight savings of 15–30 per cent. This weight reduction has led to greater fuel economy.

Kevlar is also widely used in the production of protective clothing and body armour for the police and armed forces. Kevlar is stronger than the more conventional carbon fibres, but it has greater strength and flexibility compared with existing materials. As a result, effective body protection and ballistic protective applications for military purposes are now much lighter. Police protective vests made using Kevlar are easily worn and concealed and weigh less than 2 kg, making it possible for them to be worn for many hours.

The Formula 1 racing industry has pushed forward the development of materials technology. Carbon fibre composites make this car lighter and faster

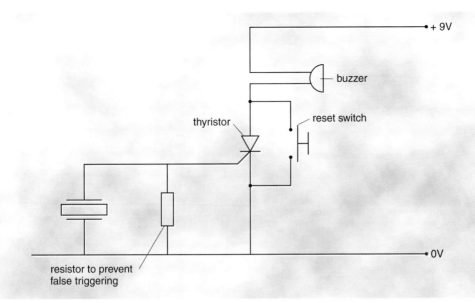

A simple circuit

Piezo crystals

Piezo electric crystals are used to make small electronic transducers. A transducer is a form of electronic output device. Other transducers include bulbs, buzzers and LEDs.

Piezo electric transducers can be used in two different ways. They can be used to produce sounds from an electrical signal. In this mode, they are used as the 'loudspeakers' in musical greetings cards.

They can also be used as electrical sensors. The transducer produces a voltage in response to a movement or loud sound. In this mode, the transducer can be used in a wide range of applications including burglar alarms.

The simple circuit shown above may act as a burglar alarm. As the transducer detects a movement or a loud sound, it produces a voltage which causes the **thyristor** to turn on an alarm. The thyristor is a type of electrical switch. However, it remains switched on even when the voltage from the transducer is zero until it is reset by pushing the reset switch.

Shape memory alloys

A metal which changes length when an electric current is passed through it falls into a group of materials known as smart materials. **Shape memory alloys (SMA)** is a more specialist term used to describe those materials that have a memory.

The most common SMA is an **alloy** called Nitinol. Nitinol is an alloy of nickel and titanium. By using a special heat treatment process, a piece of SMA can be made to remember that it should be straight at 70ºC and above.

If you take a piece of SMA wire and bend it at normal room temperature, it stays bent. When it is placed into water at 70ºC or above, it will immediately straighten out. When it cools, it will stay straight until it is bent again. The temperature at which the SMA remembers its original form and changes shape is called the transition temperature.

Smart wire is a commonly available form of SMA. Some smart wires are heat treated to remember that they should shorten when heated above their transition temperature.

Because SMA has quite a high electrical resistance, the heat required can be achieved by passing an electrical current through it.

Since SMA and smart wires can be used to create mechanical movements, they can be used as:

- green house roof vents – to let hot air out
- bath mixer taps – to control valves on the hot and cold taps
- coffee makers – to open valves so that water at the right temperature drips on to the coffee.

▪ Things to do ▪

1 Disassemble a musical greetings card and look at the electronics involved. Identify the basic parts of the circuit.

2 List some other applications that the piezo electric transducer could be used for.

3 Stick two terminal blocks on to a piece of card with a piece of smart wire fixed between them. When a small current is passed through the wire, watch the card curl up!

Use of CAD/CAM

Aim

- To understand the use of CAD/CAM to produce products in quantity cheaply.

Computer controlled machines

Computer-aided design and computer-aided manufacture (**CAD/CAM**) have allowed many more products to be made quicker and cheaply in greater quantities. The introduction of such machines as computer-controlled lathes and milling machines has revolutionized the manufacturing industry.

Greater flexibility and capacity has been provided by totally flexible manufacturing systems (**FMS**). These systems have integrated computer numerically controlled (**CNC**) machines with robotic materials handling.

A student produces 3D graphics on a combined CNC milling machine and lathe in the classroom

Non-stop machining

A computer network has now replaced the traditional workforce of tool makers and machine operators. However, the workers who are involved in the

A CNC router

programming and setting up of the CNC machines are highly skilled and trained operators. FMS now operates around the clock for seven days a week making much more efficient use of machinery.

The very nature of a computer-controlled machine means it can run unaided and unsupervised. Once it has run through a program for the first time and finished cutting, a new piece of material can be inserted automatically and the machine will cut again and again.

This advantage has allowed for the non-stop machining of products and components and because machining can now operate for longer periods, products can be produced quickly in large quantities.

Accuracy

Accuracy is also another benefit of using CAD/CAM since items produced are identical to each other even though they are being cut in large quantities at speed.

Complicated shapes

Complicated shapes and patterns are easily cut in 2D and 3D on CNC machinery. It is also possible to use a CNC router to cut MDF. An example of this is shown in the photograph opposite.

The safety factor

With the correct extraction system to suck up the waste produced, it is much safer using CNC machinery. Carrying out the same process manually with a number of potential hazards to operators is no longer necessary.

Advantages of CAD/CAM in production

- It is accurate and can produce identical items at high speed.
- Production can run for long periods without stopping.
- It can work safely and can be used to cut dangerous materials.
- Production times are quicker.
- Quality and safety controls are built into the production and operation of the system.

■ Things to do ■

1 Why are there fewer injuries to operators when using CAM systems rather than operating machinery manually?

2 Why is CAM used to make a large number of identical items?

The impact of values and issues

Aims

- To recognize that cultural issues are an important part of designing and making food products.
- To understand the moral issues of design.
- To understand changing fashions and planned product obsolescence.

Moral issues

A consumer driven society

Our society is consumer driven by new products and technologies and by material goods that clever marketing makes us believe we must have. Products are designed and manufactured for mass markets around the world.

Widespread appeal

Mobile phones are a classic example of this 'must have' marketing philosophy by huge global companies. The market has now almost reached saturation point with well over 30 million mobile phones in the UK alone.

The marketing behind such products involves developing widespread appeal to all types of customers – the young, the old, the professional, and so on. This type of organization relies heavily upon selling large quantities of its products at relatively small profit margins.

One way of increasing profit margins is for companies to manufacture their products in developing countries where labour costs and overhead costs are less expensive.

Clever marketing of mobile telephones has been aimed at young people

The Baygen radio – Aimed at developing countries with scarce energy resources, the power source is an integral spring-driven generator powered by hand. Winding the handle for 25 seconds gives over 30 minutes of play time

The Baygen radio

The problem – lack of access to information

The radio has long been used as a method of communicating news and information all around the world. Difficulties over the supply of electricity and batteries in certain regions of the world have restricted some African countries from having access to such communication systems.

The product solution

The inventor Trevor Bayliss looked at this problem from the viewpoint of wanting to provide health education and information via the BBC World Service broadcast on the radio. As a result, the 'Baygen' radio was developed. This radio uses both a solar cell for power along with a wind-up clockwork spring mechanism.

In African countries, the radios are used in education reform programmes to try to stem the spread of HIV and AIDS. This is an excellent example of a product that has been developed to tackle real moral and health issues.

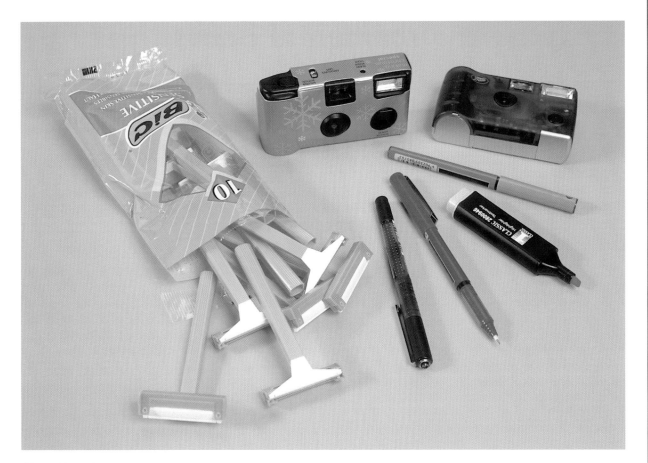

Disposable products

Product life

Products are designed to meet a specific need or requirement and designers and manufacturers have to ensure that they are fit for the intended use. This places some responsibility on the designer to make sure that the product will perform safely and reliably during its intended life span.

The life expectancy of many products can be prolonged or maintained with regular upkeep and planned maintenance. Many products, such as window frames, bridges, road surfaces, cars, aircraft, should all be regularly serviced to ensure they operate and perform efficiently, effectively and safely.

Disposable products

On the other hand, some products are designed to be used only once or for just a short period of time. For example, ballpoint pens, disposable razors and single-use cameras are all made in high volume and thrown away and disposed of when the ink runs out, the blade is blunt or the film is finished. This is called planned obsolescence.

It is in the interests of designers and manufacturers to develop and promote disposable products because they will increase their turnover and profits. This is not necessarily the best solution, however, as far as the environment is concerned.

All plastic products and components are derived from oil. As a natural resource, oil is running out and will not be available forever. It is also very difficult to dispose of plastics by either burning them or putting them into landfill sites.

Throw-away philosophy

In today's society, however, we tend to have a throw-away philosophy. If something does not work or is broken, we throw it away. Engineers sent to repair heating boilers and items like washing machines often simply remove and replace the whole electronic circuit board. This saves time since the circuits are now so complicated that it would take a long time for an engineer to diagnose and repair the fault. The faulty board is removed, thrown away and a new one inserted.

■ Things to do ■

1 Make a list of disposable products that you use and throw away when they have either run out, worn out or broken.

2 Choose one item from the products in the picture above and make some suggestions as to how the life of the product could be extended.

Environmental issues

Aim

- To understand the environmental issues that affect the design, manufacture and disposal of products.

Demand on resources

As manufacturing industries have increased output to meet consumer demands and needs, there has been an ever increasing demand on the world's natural resources. Most of these resources are non-renewable such as oil and others like timber take years to grow.

This is especially true of hardwoods which take hundreds of years to reach maturity. Softwoods mature much quicker and can be felled and taken to the sawmill ready for processing. In managed forest areas, softwoods can reach full maturity within 30 years.

The construction industry uses a lot of softwood for roof trusses, floor joists, staircases, floorboards and internal joinery like skirting boards and door frames. The industry has made some moves towards replacing floorboards with chipboard and using skirting board and architrave made from MDF.

Pollution

As manufacturing has increased, so has the waste and pollution generated. Fumes and gases given off from coal burning fire stations contribute to 'acid' rain. This has damaged the environment by causing decay to brickwork and polluting lakes and forest areas. The European Commission and the UK government have introduced very tough controls to make factories reduce their emissions.

Carbon monoxide which has contributed to the 'greenhouse' effect has also made governments worldwide look at how they might be able to reduce their levels of emissions. Fumes from cars and factories have been cut in a move to reduce the long-term effect of the greenhouse effect and global warming.

In factories where harmful fumes and dust are produced, careful consideration has to be given to how they are controlled. Control of Substances Hazardous to Health (COSHH) data is analysed and risk assessments are made. Strategies are put in place to reduce the risk and to minimize the effects on any workers and on the surrounding areas and environment.

Conservation of resources

Any manufacturing involves a certain amount of energy. This energy comes from sources such as oil, gas and coal burning power stations. These sources of energy have a finite supply and are running out quickly. Alternative methods will need to be found and much more efficient designs and production methods will also have to be adopted.

One attempt to reduce the amount of energy used in the steel industry has been to increase the use of recycled steel in the production of new steel. This greatly reduces the energy required since the old steel has simply to be re-melted in a furnace and not made from scratch in a blast furnace where much of the energy is used.

The construction industry uses large quantities of softwood

Overuse of timbers results in deforestation

The term resources also extends to the materials themselves. Essentially, all plastics are derived from oil in one form or another. In a world which is running out of natural resources, designers are being forced to consider alternative materials and solutions to problems. They have had to come to terms with these difficulties and have had to look at different strategies to overcome the demands on the environment. Some examples are:

• to use natural materials from sustainable and managed sources
• to use biodegradable materials, especially for packaging
• to increase the use of recycled materials
• to design energy saving products with increased life expectancy.

In an attempt to reduce the amount of timber needed in the construction industry, engineers have come up with an alternative metal floor joist. Not only is it easier for the plumber and the electrician to install their services, but it has also eliminated the need for huge sections of timber. The problem though is still one of economics: what is more cost-effective (cheaper) to use – a tree or a sectioned metal joist?

Waste management

The amount of waste generated by the average house in a week is unbelievable. The majority of it is packaging. Attempts have been made by some local councils and authorities to promote a kerb side collection policy. Houses are provided with separate storage bins for items such as paper and card, glass, plastic and garden waste. All of the rubbish is then collected in separate bins and recycled.

Supermarkets and some town car parks provide recycling areas too, but the onus is very much on the individual in these circumstances. The use of recycled materials is becoming much more evident in products on the high street. A great deal of packaging is now made out of recycled paper and card, and greetings cards and writing paper are branded as being made from recycled materials.

Plastics present problems in relation to recycling. There are, however, biodegradable plastics now finding their way into the market for uses such as food packaging and 'green' credit cards.

Environmentally friendly plastics such as 'Biopol' are made by fermenting sugar and natural carbohydrates extracted from vegetables. Once processed and turned into a product, they can be disposed of by chopping them up and throwing them on to a compost heap where they biodegrade.

The emphasis on designers and engineers today is certainly to reduce, reuse and to recycle. How the product will be disposed of when it reaches the end of its useful working life has to be considered.

▪ Things to do ▪

1 Look at the various parts and materials used in a disposable pen. Explain how the various parts could be seperated and recycled.

2 Find out what is recycled in your school and how you could improve it.

3 Look at several different food products from a supermarket. How could the packaging be reduced?

Influences of different cultures

Aesthetically pleasing and appropriate for users

Designers are very concerned with the **aesthetics** of their products – how they look. However, what is aesthetically pleasing to one person is not necessarily aesthetically pleasing to another. Another issue the designer must consider is whether the product is appropriate for where it is going to be used and for the group of users.

Different views

People of different cultures have different views and opinions on what is aesthetically pleasing. Often these views are based on historical and religious images and works. It is, therefore, important for designers to be aware of both the cultural and artistic influences on their work as well as the overall design for manufacture.

Issues to consider

This leads to a number of issues such as:

- the availability of materials and skilled labour
- the physical resources such as plant and machinery
- the environmental impact – whether the product damages or improves the local environment.

Each chair is from a different period in history and from a different part of the world. Each one reflects the cultural aspects of its country of origin and an artistic interpretation of the materials and manufacturing capability of the time. Another consideration is whether the chairs were made in times of little or no automatic machines. The emphasis would be, therefore, on the craftsman and his ability to be able to translate the designs into reality.

Image

Many consumers, often young consumers, are concerned with the image that a product gives them. With this in mind, designers are faced by increasing pressures to create products which give consumers an image. What is created and the image it gives out can be termed as a cultural element.

Mobile phones and changing status

Fashion accessories and clothing are very closely associated with different cultural groups in today's society. In the 1980s, when mobile phones first

These late 19th century Prie-dieu chairs of India are made of carved blackwood

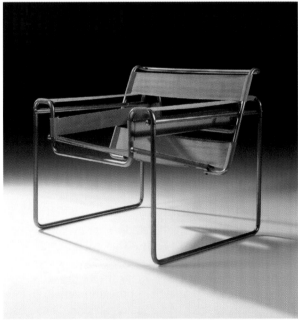

The Wassily chair, made of tubular steel, was designed by Marcel Breuer, 1925

Mobile phone from the 1980s

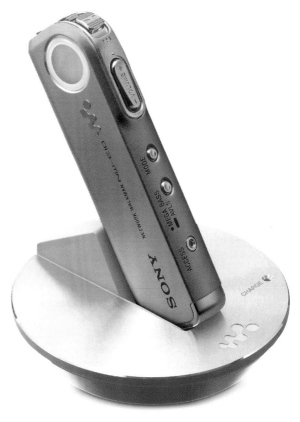

became commercially available, they were widely associated with a group of people known as 'yuppies' – young upwardly mobile professionals. These phones were regarded as a sign of wealth and professional status.

Today, mobile phones are commonplace and are changed almost as often as shoes. The design and marketing of products has influenced how they are perceived and need to adapt and change as the image of the product changes.

From Sony Walkman to MP3

Music often seems to create and reflect cultural trends. Products such as the Sony Walkman set the trend in providing people with the ability to listen to music on the move. As technology developed and the power of electronics increased, miniaturization allowed products to become smaller and more compact.

The latest personal miniature stereo system

Future trends

As designers create new products, they must be aware of current and future trends and market needs. They must also consider the cultural, social, moral and environmental issues in relation to the product and where it is going to be used in the world.

■ Things to do ■

1 For each of the chairs shown on the previous page, work out what tools and processes may have been used in their construction.

2 Select a range of desk lights from different periods of history. Look at them and think about whether they were designed for manufacture or for their aesthetic appeal.

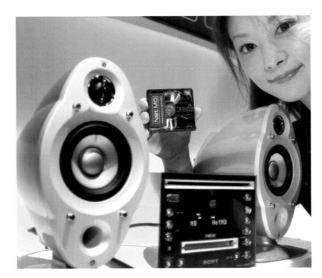

Japanese companies have produced many innovative miniaturized products

Product analysis

Aims

- To understand the need to analyse existing products.

Criteria for product analysis

Before a designer begins to design a new product he or she usually conducts a product analysis of similar products or production techniques that already exist. This strategy is especially important if the main aim of the new design is to improve upon an existing product in terms of either the aesthetics or the performance. It also helps to ensure that a new design is fit for its intended purpose, that is, it does what it is supposed to do.

Analysing products takes time and practice. It is not simply a question of saying how much you like them. The process can be focused around a series of questions. When analysing a product these are the sorts of questions you should be asking:

- What is the aim of the product?
- What should it do and how is it made?
- Who is it for – target group?
- Why will they need it or want to use it?
- Where will it be used – the type of environment?
- How does it look?
- How does it work?
- What materials is it made with?
- What standard components and fixings have been used?
- What manufacturing processes have been used?
- What quality and safety aspects does it have?
- How well does it do what it is intended to do?
- How long will it last for and what maintenance is involved?
- What environmental issues are involved?

Quality of design and manufacture are both important aspects of any design since no product will sell well if it is of poor quality. Two points are normally considered when buying a new product:

- Does it look good (**aesthetics**)?
- Does it work well (function)?

These questions lead to a common debate about good design. Should form, or the product's aesthetic appeal, be put before its function, or vice versa?

Form versus function

The kettle shown here is a classic example of the debate about form versus function. The kettle as a basic product exists in almost every household. Its function is simple in that it boils water. In the case of the Hot Bertaa, by Philippe Starck, the design pushed the form versus function argument to the extreme – that is, its form was considered without too much regard to its function. The packaging on the product stated 'Do not use when hot'. Quite a problem for a kettle! The design now stands as an icon of the post-modernist trend during the 1980s.

We can also spot cultural differences in the types of kettles that are used around the world. For example, most kettles in the UK are of the electric type, whereas hob kettles are widely used in Scandinavia, France and Italy. With this in mind, designers and manufacturers alike must survey where in the world the product is to be used and what implications this has. They can gather this information through analysing similar products in the market they are targeting.

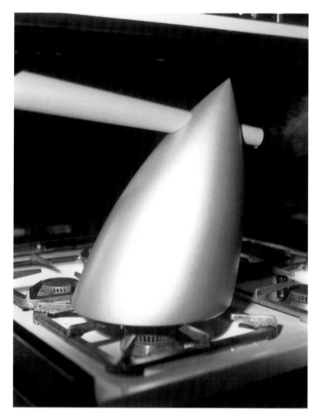

Hot Bertaa kettle by Starck

Disassembly

Another strategy used in product analysis is to disassemble items. It is easy to dismantle a product to see:

- how it has been manufactured
- what components have been used
- how it works
- what materials and processes were involved.

Recording observations

In an attempt to formally record any observations made, it is sometimes easier to make a table like the one below. This is especially helpful when it comes to analysing the individual parts of a product.

Value for money

The wine rack shown below is sold in a cheap flat-packed form. One question that you should always ask of a mass-produced product such as this is 'Does it represent good value for money?' On investigation, you would be able to look closely at the quality of materials and finish and judge for yourself.

▪ Things to do ▪

1 Collect a range of simple products that you think have been designed well. Analyse these products using the questions listed under 'Criteria for product analysis'.

2 Choose one product from your collection. Describe the manufacturing processes used to make it and how well the separate pieces fit together.

Part	Quantity	Material	Finish	Process
Support pegs	62	Birch dowel	None	Drawn through a die
Corner supports	40	Pine	None	Rough sawn, planned and drilled

Recorded observations

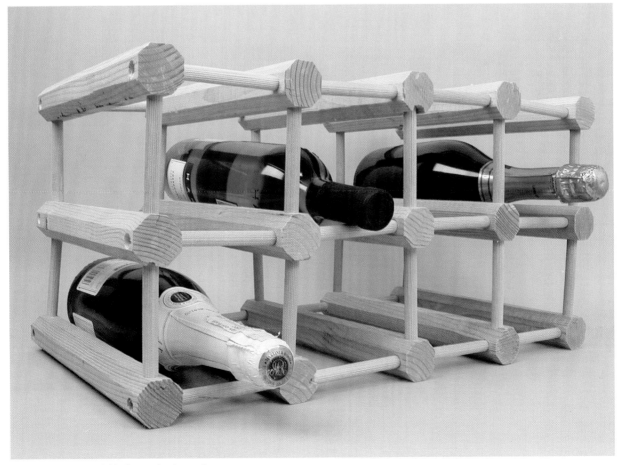

A commercially available flat pack wine rack

Design for production

Aim
- To understand how important it is to design products with production in mind.

The ideal product design can only exist if the designer creates a product that is:

- functional
- aesthetically pleasing
- financially viable
- able to be manufactured.

This 'total' concept of design with the consumer, function, cost and manufacture all given equal consideration is what is meant by design for production.

It is however unrealistic to expect a single designer to be a specialist in each one of these areas. Because of this, designers often work in association with manufacturers' production departments, who are specialists in manufacturing technologies. It is their responsibility to suggest and make recommendations to the designer or design team in order to make production more efficient and more cost-effective.

It is essential that the design for production philosophy is adopted at the earliest possible stage in the design process. There will inevitably be conflicts that arise along the way, but any conflict that arises has to be resolved in order to reach a compromise and get a successful product into the market place.

However skilled and creative designers are, they will usually try to make full use of standard materials and standard parts such as gears and screws. They will also try to make use of standard machines and machine tools rather than having to invest money into new specific machinery. This process is known as standardization.

It is of great economic benefit to make use of standard parts, especially across a range of products such as cars, for example. This means that only one machine needs to be tooled up in terms of moulds to

Spot the difference!
When you see two cars from the same range parked next to each other in a car park, try and spot some of the similarities and differences.

Ford Mondeo interiors from the 'X' and 'Ghia' series. The main instrument panel mouldings are common components. Series differentiation is achieved by varying selected sub-components

be able to stamp out the individual pieces that go to make up the car itself. Since the machines cost millions of pounds to buy and the tools hundreds of thousands of pounds to set up, the manufacturer cannot afford to produce a range of cars each having different wings and door panels. These have to be standardized to make the range cost-effective.

However, on cars across a product range many varied components are also used to give different models a specific look. Among those that we can easily spot are locks, door handles, wing mirrors, steering wheels, dashboard layouts, lamps, indicator covers, radios, gear sticks, seats and many more.

When creating a design, after conducting a product analysis the designer will produce some initial ideas. You can see from the initial design drawings of a CD rack how the shapes and ideas differ. In these early stages it is important to introduce as many ideas and variations as possible. Once the ideas are complete, discussions can take place between the design team

and the manufacturers. At this point, decisions would be taken about which ideas to develop through to a final idea for production.

Models and working prototypes will follow as the product is further developed for production. The final decisions rest with the production engineers who are responsible for the manufacture of the product.

The decisions that face the production engineers are how to realize the design and how to do so as quickly and as cheaply as possible. They would also look at whether it was viable to injection mould several items at a time to increase production rates and overall efficiency.

■ Things to do ■

1 What is meant by the term standardization? Give some examples.

2 Why is compromise so important when it comes to designing products for manufacture?

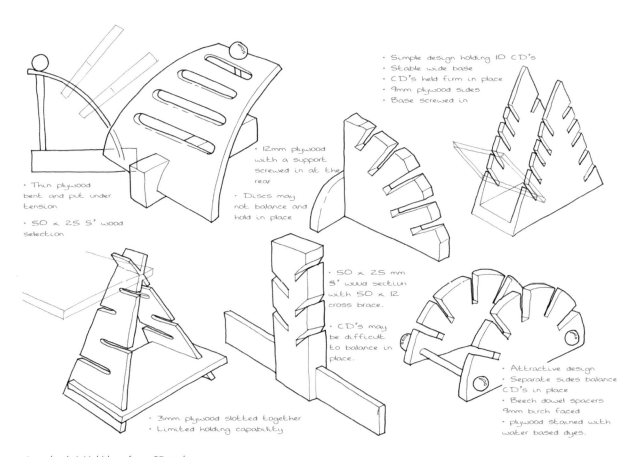

A student's initial ideas for a CD rack

Planning for production

Aims

- To understand the need for planning for production.
- To understand the terms control systems and quality control.

Many manufacturing processes exist, but the main categories involved in making a product can be broken down into six areas:

1 Materials preparation – for example sawing and planing of wood; cutting and preliminary heat treatment such as annealing.
2 Machining – for example turning, milling of metals; routing and shaping of wood.
3 Heat treatments – for example casting metals; hardening and tempering; **injection** or **blow moulding** plastics.
4 **Fabrication** – for example cutting joints in wood; riveting, welding and brazing of metals.
5 Assembly – for example glueing woods; welding/soldering metals; use of nails, screws, nuts and bolts.
6 Finishing – for example polishing; plating; spraying.

Key stages in manufacturing

When decisions have been made about which processes are going to be used and how many items are to be made and by when, a work plan of **flowcharts** is drawn up.

Most companies now use a computer program to produce these charts and this helps them to order materials and components and estimate how long each stage will take. The program will also build in the appropriate time and place for quality control checks to take place.

Control systems

Manufacturing industries benefit enormously from control systems in:

- materials handling – when to move materials and components from one place to another and actually moving them
- preparing materials – control systems operating temperature-controlled ovens for heat treatment, computer-controlled machines for cutting and stamping materials
- processing operations – automatic welding and checking of welds.

In almost all cases, control systems are electronic with dedicated systems and microcomputer control devices. Mechanical devices such as solenoids and motors, and pneumatic and hydraulic systems often do the work.

Control systems have many advantages in manufacturing:

- They can control machines and devices to carry out the same function or process time after time in exactly the same way.
- They can carry out processes to high levels of accuracy and **tolerances**.
- Automated systems can be used for tedious and repetitive tasks.
- They can increase speed of production and increase productivity.
- They can be used to control devices in dangerous and hostile environments.
- They reduce the risk of human error.
- They alert the need for repair and maintenance through the use of warning and safety systems.

Control systems can be used in:

- production and assembly lines for materials processing and handling
- processing operations
- **batch** and high-volume **production** runs
- flexible manufacturing systems (**FMS**)
- **quality control** systems.

Quality control

One of the biggest challenges facing the manufacturing industry today is that of quality. The specification is a key part in the whole **quality assurance** process. The product specification is used to describe and detail each part of the product. It will include:

- the material which is to be used and its **mechanical properties**
- any form of heat treatment or processing
- its overall dimensions
- its tolerance for manufacture.

The **mass production** of any product relies on a degree of interchangeability. This means, for example, that any shaft made must be able to fit into any bearing case. If all shafts fall within the tolerance limits, then unit costs of production are reduced.

Turning a single shaft in school on a centre lathe involves taking a number of small cuts to gradually reduce the diameter. After each cut a check should be made with a micrometer or vernier caliper until the correct size is reached. This is a slow process and cannot be used in industry to produce shafts in batches or high volume.

In industry, a machine would be set up specifically to remove a fixed amount of material to leave the required size. The only problem is that the machine tool wears down and the shaft diameter gets very slightly larger as the machine works. This is another reason for having a tolerance limit on the component.

The checking of the shafts becomes a quality control issue at this stage. Once a batch of 50 is completed, for example, a quality control inspector will take a sample of shafts at random from the batch. This is known as sampling and the inspector will check each one of the sample to ensure that it is within the tolerance limit.

Sampling is often carried out using specifically designed and made gauges known as go/ no-go gauges. These would be set to the upper and lower tolerance limit. Any which are too big or too small are rejected at this stage.

■ Things to do ■

1 What are the main advantages of using control systems?

2 Explain what is meant by the term sampling with reference to quality control checks.

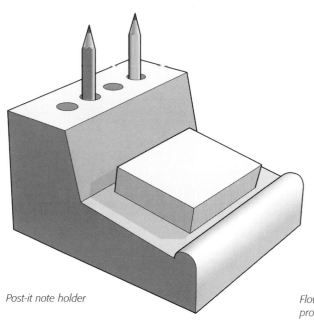

Post-it note holder

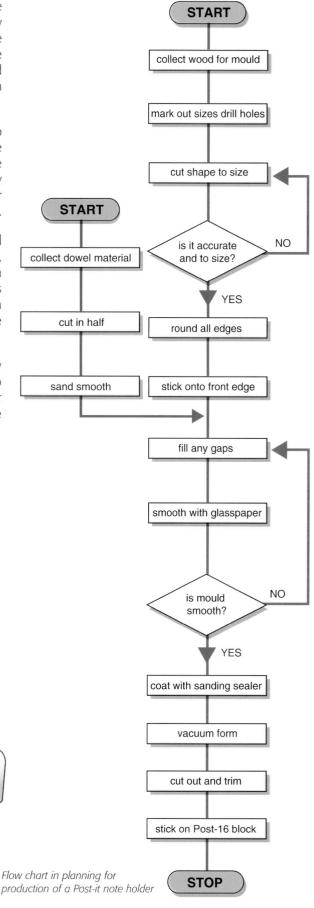

Flow chart in planning for production of a Post-it note holder

Design and market influences

One-off design

These pages describe the work of a student who was involved in designing and making a table for a GPs' reception area.

Case study: GPs' waiting-room table

Waiting rooms at doctors' surgeries, dentists, vets and hospitals can often be quite solemn looking places. The decoration is generally bland with uncomfortable chairs and a few old magazines scattered on an old coffee table.

In consultation with the doctors at one surgery, a proposal was put forward to design a new table as a focal point for the waiting room. This type of design and manufacture falls into the one-off category.

Initially, a brainstorming exercise took place to come up with a range of possible themes to be explored, and areas like medical instruments were looked at with a view to basing the design around them.

The main emphasis of the design had to reflect the nature of the doctors' business. After all the initial ideas had been presented to the client, it was decided to develop a syringe idea.

Much of the development work centred around what sizes of material were available from various suppliers. The chosen material was acrylic for the tabletop and the legs. Ideally, glass would have been used but in the prototyping stages glass proved to be too expensive. It needed to be drilled in 16 places and the edges polished and tempered for safety reasons. Acrylic for the model syringes was an appropriate choice of material. Although acrylic is a brittle material, it is fairly hard and in a round tube section it is rigid and stands up well to **torsional forces.**

Various small models were made of syringes after careful measurements were taken from a real syringe and scaled up to make sure that the proportions were maintained. The nearest available sections were chosen from the manufacturer's catalogue and working drawings produced enabling the manufacturing processes to start.

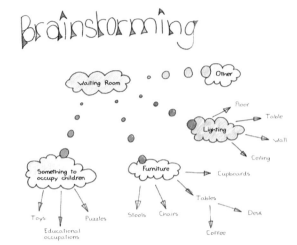

Brainstorming

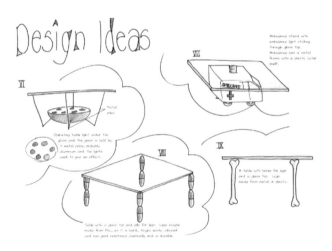

Design ideas

Further design ideas

Even though the table was designed as a one-off, appropriate use was made of the **CAD/CAM** system. CAD/CAM is very efficient when it comes to repeatedly cutting out identical components. Although there are several pieces to each leg, they are identical and quite complicated. CAD/CAM was the best solution for the manufacture of the separate pieces and once they had been drawn on to the CAD package, they were easily cut on the CAM machine. The accuracy of the machine made the fitting of the parts together easier. In most cases, the various parts were held together by an interference fit. Where bits did need to be assembled they were glued together using Tensol cement.

Ideally, the black tips would have been cut on a CNC lathe. Unfortunately, one was not available so they were cut manually on a traditional centre lathe. All items that were cut on the CAD/CAM system had their edges polished on a buffing wheel. The top plunger part of the syringe had to have the four holes in it tapped to take an M8 countersunk stainless steel machine screw. Each leg was held on to the table top with four of these screws.

The completed table was taken to the doctors' surgery and left in the waiting room to be tested. It was always going to be subjected to bangs and knocks but it stood up to the test very well. Obviously, the tabletop itself became a little scratched, but this was because it was made from acrylic and not glass.

The table largely received a warm welcome from patients during the testing period, although some people found the use of syringes for legs off-putting.

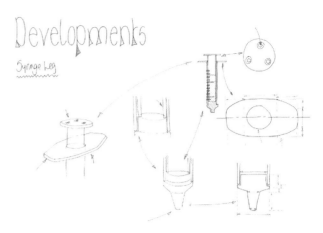

Developments

There was also a concern that the table wobbled slightly when knocked.

As a one-off design, the table is a unique piece. The product generally fulfilled its function and it performed reasonably well in its testing period. A number of minor changes and modifications were suggested and recorded in an attempt to improve upon the initial design and these have been taken on board by the designer as genuine improvements.

■ Things to do ■

1   Look at some previous projects which have been completed in your school. Analyse and evaluate them and investigate the processes involved in their manufacture.

Finished product

Batch production

Aim

- To understand the design, consultation, planning, evaluating and manufacture involved in batch production.

These pages describe the work of a designer who was involved in designing and making some plaques to commemorate a school cricket club's achievements.

Cricket plaques

A school cricket club with a long-standing tradition decided it wanted to honour some of its finest players and achievements on the cricket pitch. Following discussions between current and past players, it decided to display a plaque for those players who scored either a century (100 runs) or who took seven wickets in a single innings.

Having checked back through all the old score books and archives, the club found that it needed 31 plaques, spanning a history of more than 70 years. The plaques were to be planted around trees in the grounds of the school.

The members of the cricket club were clear about what they wanted the plaques to look like and what materials they wanted them to be made from. The base of the plaque was to be made from oak, an English hardwood. It was to be fixed into the ground by means of a metal post concreted in. The actual plaque was initially requested to be in brass.

Production methods

Batch production was going to be the most appropriate production method to use, even though the actual inscribed plaque carrying the person's name and record was going to be produced as a one-off. Nevertheless, at the end of each season, another batch of plaques was going to be needed to celebrate that season's success.

Requirements

The club members decided that the plaque had to:

- look professional
- stand up to all types of weather conditions
- remain in the ground
- fit in with very traditional surroundings
- offer value for money
- last a long time
- be visible and obvious to those walking around the grounds.

Materials

The use of oak was a suitable choice. Providing it was correctly finished with an exterior grade polyurethane-type varnish, it would stand up to many years outside. An optimum size was reached in consultation with the school's technician in relation to the available widths of oak from the timber supplier.

In terms of the metal spike used to anchor it into the ground, a number of possibilities were discussed. Finally, an angle iron was chosen because it provided a flat fixing surface for the rear of the plaque, and when cut and welded in position, was stiff and rigid enough to support the plaque without flexing.

Making

A series of **jigs** and **templates** was designed and made to aid the production in the marking out stages. A template was also used to mark out the small spike.

The oak back plate was cut to the approximate size. It was then run through the planner thicknesser to reduce it to the final thickness. Next, it was cut into the appropriate lengths on the circular saw.

From this point, the oak was hand machined on the router to produce a decorative edge. For safety purposes, the router was turned upside down and held in a special table. As the blade rotates in the router and the wood is passed over it, it machines a profile on the edge of the timber.

Initial artist's impression

The proposed solutions

The choice of screws that would fix the angle iron spike to the oak plaque was carefully considered. Steel screws will react with oak and eventually corrode. Therefore, brass screws were used to secure the two parts together although an identical sized steel screw was first inserted to make the correct sized hole. This is because a brass screw is not hard enough to cut into the oak and the head would be damaged.

The inscribed plaque underwent some development where a number of alternative methods were made and evaluated. They were judged against their looks, cost and production times in relation to value for money.

Alternative solutions were proposed and made. Each plaque had to contain a certain amount of information and so choice of font and size of font was crucial. The solutions were:

- a **CNC** engraved plastic laminate sheet
- a special 'sublijet' printing technique on to a polyester coated brush brass sheet.

After much deliberation, the 'sublijet' printing process was chosen. The whole process was very quick and cost-effective and the sheet could also be cut very cleanly on a guillotine, making for a quick and efficient process.

The brass sheet was fixed to the oak back with brass round head screws for a very professional finish. The plaques were completed in one large batch and then put in place around the trees. At the end of each subsequent cricket season, a new batch of plaques was manufactured.

▪ Things to do ▪

1 Consider how a product you have designed and made at school might have needed to be changed if it had to be batch produced in hundreds.

Aim

● To understand the issues related to the design of a product which is to be produced in high volumes.

Case study: IKEA's IVAR shelving range

One product that has been successfully designed for high volume production is IKEA's IVAR shelving range. It is modular, meaning you can create different pieces of furniture from its main parts in the range. This case study looks at some of the features of the design which has led to the production of its component parts in high volume, yet still allowing freedom for consumers to create and construct their own individual piece of furniture.

IKEA has now established itself as a market leader when it comes to home furnishings. Much of its furniture is sold in flat-pack form to be assembled at home by the user.

Flat-pack furniture has many advantages for the store and the user:

* prices are lower because there is no final assembly involved for the manufacturer
* it takes up less space in the warehouse
* the consumer can generally take it home in the car
* a variety of units allows consumers to create their own designs to suit their own home and environment

Environmentally friendly policy

IKEA has adopted an environmentally friendly policy as far as its use of timber is concerned. The company refuses to use solid wood originating from intact natural forest. The IVAR shelving range is left untreated and natural. Customers can colour the material with paint, varnish or dye if required.

Adaptability

This type of shelving system can be adapted to suit all needs. Within the range, there are four different

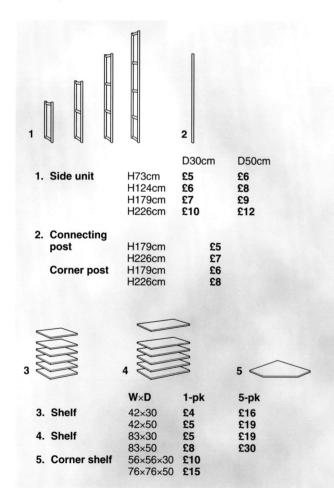

		D30cm	D50cm
1. Side unit	H73cm	£5	£6
	H124cm	£6	£8
	H179cm	£7	£9
	H226cm	£10	£12
2. Connecting post	H179cm	£5	
	H226cm	£7	
Corner post	H179cm	£6	
	H226cm	£8	

	W×D	1-pk	5-pk
3. Shelf	42×30	£4	£16
	42×50	£5	£19
4. Shelf	83×30	£5	£19
	83×50	£8	£30
5. Corner shelf	56×56×30	£10	
	76×76×50	£15	

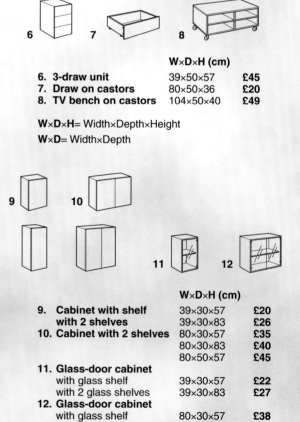

	W×D×H (cm)	
6. 3-draw unit	39×50×57	£45
7. Draw on castors	80×50×36	£20
8. TV bench on castors	104×50×40	£49

W×D×H= Width×Depth×Height
W×D= Width×Depth

	W×D×H (cm)	
9. Cabinet with shelf	39×30×57	£20
with 2 shelves	39×30×83	£26
10. Cabinet with 2 shelves	80×30×57	£35
	80×30×83	£40
	80×50×57	£45
11. Glass-door cabinet		
with glass shelf	39×30×57	£22
with 2 glass shelves	39×30×83	£27
12. Glass-door cabinet		
with glass shelf	80×30×57	£38
with 2 glass shelves	80×30×83	£43
with glass shelf	80×50×57	£47

IKEA's IVAR shelving system

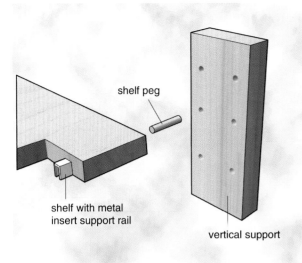

Shelf support

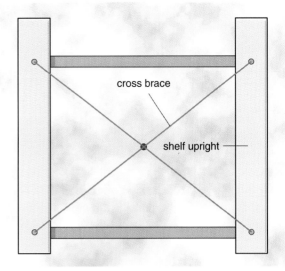

A cross brace provides stability

heights, two depths and two widths of items available. This means that you can design your own units to create a storage or display system that fits your own needs and just about any available space. The range also includes drawers, cabinets and shelves. Extras such as lighting, handles and knobs can be added to create your own individual look.

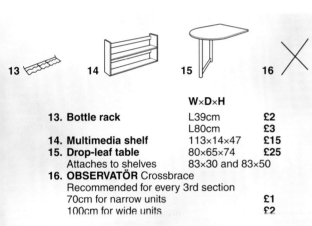

	W×D×H	
13. Bottle rack	L39cm	£2
	L80cm	£3
14. Multimedia shelf	113×14×47	£15
15. Drop-leaf table	80×65×74	£25
Attaches to shelves	83×30 and 83×50	
16. OBSERVATÖR Crossbrace		
Recommended for every 3rd section		
70cm for narrow units		£1
100cm for wide units		£2

Materials and components

The materials used in this range are simple and relatively cheap. Extensive use is made of softwood with MDF used as drawer bases. The actual softwood is of a good quality and is seasoned well to suit the temperatures and humidity levels in modern centrally heated houses. In terms of the overall safety and stability of the structure, a cross brace is used to provide some triangulation and rigidity.

The vertical uprights are drilled along their entire length so that the user has the flexibility of being able to adjust the height of the shelves. The shelves have a metal insert fixed into them across their width. This is because they are recessed to locate into the short metal studs and, therefore, would become weak across the grain and likely to break very easily.

The market

The type of people likely to buy and put together this sort of furniture tend to be the middle-aged and the younger generation. The furniture is seen as modern and could even be described as quite minimalist. Some people take great satisfaction in constructing such pieces of furniture, especially when they have 'designed' and 'made it' themselves.

This range provides great flexibility and very good value for money for a solid timber product. The ability to be able to add your own choice of colour and finish adds to its overall appeal for many customers.

■ Things to do ■

1 Explain the benefits of a modular system such as this for the consumer and retailer.

High-volume products 2

Aim

- To further understand the issues associated with and related to the design of products for high-volume production.

Case study – the disposable pen

Before the ballpoint pen was invented by two brothers called Biro in 1938, most people wrote with fountain pens. Nowadays, the ballpoint pen has largely taken over from the traditional fountain pen.

The cheapest form of a disposable Biro can be bought for around 15p. It is made up of six separate items plus the ink. The six individual pieces are shown in the diagram below. They are :

- cap
- ball bearing
- nib
- ink tube
- barrel
- end cap.

Materials and processes

Different plastics are used for the cap and end cap, ink tube and the barrel. The barrel is quite stiff, whereas the end cap, cap and ink tube are quite soft and flexible.

Barrel, end cap and cap

The barrel, end cap and cap are all made by **injection moulding**. Although the initial costs of the moulds are high, the unit cost for each individual item made is very small when they are made in large numbers.

One advantage of using a technique like this is that the final colour of a specific item can be changed easily. One colour end cap would be produced for a while with a matching cap. When enough stock has been made, the coloured granules being fed into the machine are changed and another colour inserted.

Ink tube

The ink tube is made from a much softer and more flexible type of plastic. Ink tubes are extruded, a process similar to squeezing toothpaste through a tube. **Extrusion** is an excellent process for producing uniform cross-sectional shapes. As the tube comes out of the die, it is cut or cropped to the correct length before being taken off to have the end nib and ink added.

Ball bearings

The ball bearings are made from steel which is hard and tough and the nib is made from brass which is a softer material. The ball bearings are made by feeding steel wire in between a special set of skewed rollers. These roll the wire into shape and eventually cut off the rough balls or spheres of steel. The rough balls are then fed into a high-speed grinding mechanism which gives them a very highly ground surface finish at a very high tolerance in terms of accuracy.

The ball bearing is then fitted into the brass nib with interference fit. This means that it is gripped and held in place by the brass nib, yet it is free to rotate within it allowing the ink to flow around the edges.

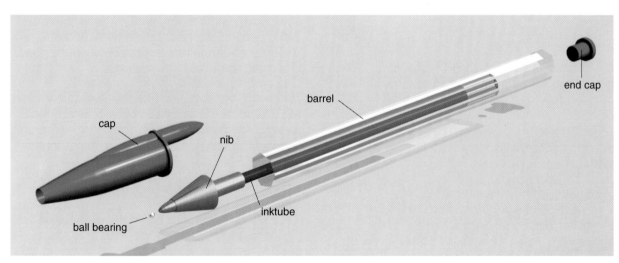

Exploded view of a Biro

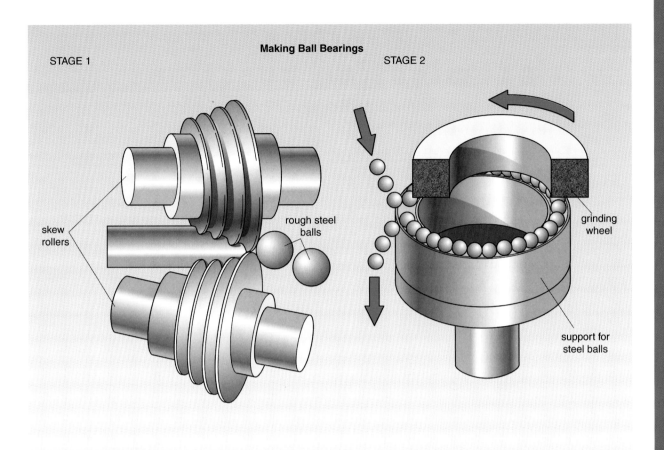

Making Ball Bearings

STAGE 1

skew rollers

rough steel balls

STAGE 2

grinding wheel

support for steel balls

Skew rollers making ball bearings

Ink

One of the difficulties in creating the Biro was in identifying the best ink to use. Normal ink was too runny and would leak out around the bearing, so a muck thicker ink is used, which is more **viscous**. Although one end of the ink tube is open, the ink will not run out if the pen is tipped upside down. This happens not just because the ink is more viscous, but also because the surface tension forms a **meniscus** on the exposed surface and because the tube is small there is not enough weight in the ink tube to break the surface tension.

Recent developments

In recent years, a small number of changes have been made to the design of the Biro. These followed a number of accidents where children swallowed the cap and choked. The cap now has a hole in the end so that in the event of a child swallowing it, the child will still be able to breath since air can pass through the cap.

Many types of disposable pen now exist. Different shaped barrels are used and new and improved ergonomic features have been added to improve grip and comfort. As some have retractable ink tubes, the need for caps has been removed.

Environmental consideration

When a basic ballpoint runs out of ink, it is useless and has reached the end of its expected life cycle. As our valuable resources are running out and all of the plastic components used in disposable pens are derived from oil, maybe we should give more thought to what we do with these items or how we could extend their life.

After all, if we buy a more expensive ballpoint pen, we can buy refills when the ink runs out. Should manufacturers of basic ballpoints consider selling new ink tubes to replace old ones? We would then be saving the end cap, cap and barrel.

▪ Things to do ▪

1   Collect a range of disposable pens and list the differences between them.

2   Which is the most comfortable type of pen to use? Carry out some consumer tests.

1 The drawing below shows a desk letter rack and storage unit for pens and paperclips.

> Additional information:
>
> The unit is to be used on an office desk.
> Two specification points for the container are:
>
> – It must hold envelopes measuring
>   220 mm x 110 mm.
> – It must be easy to remove envelopes from
>   the unit.

a Give three more points which could be included in the specification of this product. For each point, give a reason why it should be included. **(3 marks)**

b Name the specific type of material suitable for making each of the following parts of the unit:

i wooden base

ii plastic envelope separators. **(2 marks)**

c Give one property associated with one of the materials you have named in (b) and explain how this property makes it suitable for this application. **(3 marks)**

2 Kevlar is a composite material.

a Explain what is meant by a composite material. **(2 marks)**

Lighter, stronger products such as body armour used by the police and the military are now made from Kevlar.

b Describe what is meant by the term 'strength to weight ratio'. **(2 marks)**

3 Much use is now made of CAD/CAM systems in manufacturing.

a List four advantages of using CAD/CAM systems. **(4 marks)**

b Describe two effects that built-in obsolescence would have on the design of a product and explain what impact each will have on the environment. **(4 marks)**

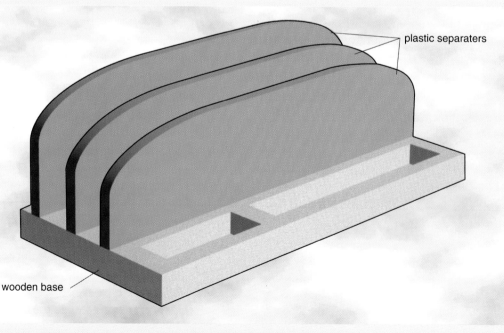

plastic separaters

wooden base

Desk tidy and letter rack

Section E
Full course coursework

Criteria 1

Aim

- To identify needs and to use information sources to develop detailed specifications and criteria.

Identifying a need

Designers design new products for a number of reasons:

- to solve a problem
- to improve the performance of an existing product which may function poorly, be unreliable or simply look out of date
- to redesign a product because a new technology has been introduced
- to improve the sales of an existing product by improved graphics or new colours and packaging.

Any of the above reasons can help you to decide upon a situation for your design project. Brainstorm some ideas for your own situation. Start with areas such as play, schools, leisure or the home. Then add in products to do with these situations, or any other areas that you may be interested in.

> **■ Hints and tips ■**
>
> Take some time to consider carefully all the projects that you have come up with. Discuss the project titles with your teacher in order to make sure that the project will allow you to complete all the assessment criteria.

Your product should fulfil a genuine need. One way of finding out about needs is by talking to different groups of people and asking what sort of problems they meet in their everyday activities.

Talk to your family, friends and other members of the community about this. Make sure that you make a record of your conversations to help you when you start to put together your design brief.

> **I To be successful you will:**
> - Identify a realistic need
> - Explore the problem through analysis and investigation
> - Produce a detailed design brief

Marks awarded: 3

Gathering information

Before starting to design and make your product, it is important to do your research thoroughly. Gather information from a wide range of sources to give you as much useful data as possible. Possible sources you could use might be:

- market research
- consumer surveys
- visits to manufacturers
- product test reports such as *Which?*
- textbooks
- e-mail
- CD-ROMs
- databases
- the Internet
- data sheets
- magazines
- people in the manufacturing industry.

You should analyse the information you collect and use it to justify the need you are trying to satisfy. You should also use it to help you make informed decisions when you reach the initial design stages of your project.

> **■ Hints and tips ■**
>
> The information you use must be relevant and targeted specifically to your project. Useless padding does not gain marks!

In industry, a lot of time and effort is spent on product research and development. Many important questions are asked such as:

- Is there a real need for the product?
- Who are the potential users?
- What is the target market group?
- Is the market large enough?
- What environment will the product be used in?
- What existing products are already available?
- What are the opinions of the users of a good and a bad product?

It is certainly worth starting with some of these points in the initial stages of your research. It might also be useful to interview somebody about your product.

> **I To be successful you will:**
> - Gather and analyse useful information which can be used to develop a detailed specification and criteria.

Marks awarded: 3

Presenting information

It is important to be selective when you present your findings. Experiment with different presentation techniques and media. Below are some examples of artwork showing different ways you can present information.

For example, information can be presented:

- visually, showing measurements
- using ICT to show the results of surveys
- as written extracts of interviews
- using symbols.

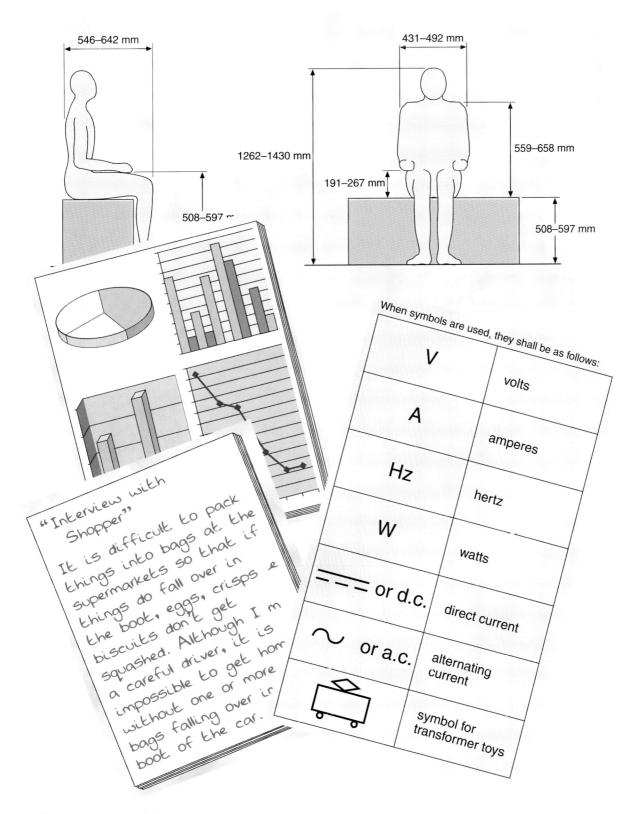

When symbols are used, they shall be as follows:

V	volts
A	amperes
Hz	hertz
W	watts
━ ━ ━ or d.c.	direct current
∿ or a.c.	alternating current
	symbol for transformer toys

"Interview with Shopper"

It is difficult to pack things into bags at the supermarkets so that if things do fall over in the boot, eggs, crisps & biscuits don't get squashed. Although I m a careful driver, it is impossible to get hom without one or more bags falling over ir boot of the car.

Different ways to present information

Design brief and specification
Producing a design brief

Once you have gathered and analysed all your information, you will need to produce a design brief. This is a short statement about what you are intending to design and make. It states the problem and provides some detail about:

- who it affects
- what happens as a result
- when it occurs
- why it is a problem.

For example, if a designer was asked to design a product to carry and display an open map on a bicycle, he or she would have to consider the following issues as part of the problem analysis:

- *Who does it affect?* The cyclist and possibly other road users
- *What happens as a result of not having this map displayed?* The cyclist has to stop, which may cause congestion.
- *When does it occur?* All the time, but mostly on cycling holidays, weekends.
- *Why is it a problem?* Safety of cyclist and other road users, inconvenience.

Stating the need and gathering extra information is the first step towards developing a design brief. As you can see from the example opposite, the situation sets the context for the design brief.

Having produced a design brief, you should now carry out any further research and analysis required. This will enable you to produce a product specification, which will guide you through the design process and product development.

Producing a specification

Designers in industry will often be given a design brief by their client. This will enable them to start extensive research and analysis of the problem. From this, they will produce a product specification that will guide them through the product's design and development.

A product specification is a list of the product's main functions and qualities. The specification should contain quantitative and qualitative information.

Quantitative information can be measured in many ways and includes details such as a maximum weight or the overall dimensions.

For your project

Once you have produced a statement about a need and collected and analysed your information, write a design brief.

Qualitative information provides details such as:

- the materials
- the purpose of the product
- scale of production
- appearance
- safety factors
- product maintenance
- environmental issues.

The specification should include statements that are specific to the product such as:

- It will be made of acrylic.
- It will have a space for storing paper and envelopes.

Your product specification needs to include enough detail to guide you in your thinking and be a basis for generating ideas. It may be that as you start designing, your specification changes. You should keep all versions of the specification to show the development and refinement of the design.

The specification should also be used when you evaluate your initial ideas.

For your project

Produce a specification that describes the following for your design:

- form
- function
- user requirements
- budgetary constraints.

To be successful you will:

- Produce a specification that describes form, function, user requirements and budgetary constraints.

Marks awarded: 3

Situation

From my research I have confirmed that whilst many people enjoy cycling holidays, they need to refer to maps.

Currently they have to stop, get out their map, read it and fold it away again. There is therefore a need for a product that can be fixed to the handlebars of a bike to display a map open whilst riding. It will have to be safely secured to the handlebars.

Design brief

To design, make, test and evaluate an attachment to fit onto a bicycle to display a map in an open position.

When and how often will the product be used?

Who will use it?

What is the design problem?

A type of solution

The need from the stated problem

What environment it will be used in

Conditions that must apply to the design

A situation and design brief showing some considerations of the design

Criteria 2

Aims

Ideas

This is a very important stage in the design process and coursework project. You are expected to generate a wide range of feasible design ideas, to present alternative solutions to a problem and to display your design ability. You should present at least three alternative designs and each design solution should meet the points of the specification. Each should be realistic and workable within your own ability and the facilities available in your school workshop.

The solutions you present as design proposals should investigate obviously different materials. Alongside each different material you should annotate (write explanatory notes) the processes and working techniques involved. This allows you the opportunity to display the knowledge and understanding you have gained throughout your Design and Technology course. The **annotation** of your work should also indicate details regarding how things work and what finishes are required.

For help with ideas, try looking at:

- natural forms – such as leaves, shells and fossils
- the work of other designers – such as Starck and Alessi
- a period of history or a design movement – such as Arts and Crafts or Bauhaus
- a type of music or a fashion
- new technologies – such as new materials like 'polymorph' or a **shape memory alloy** (**SMA**).

Here is an example of three ideas one candidate came up with for a bird feeder in three different materials.

Development of ideas

Idea development brings together the best features from the initial ideas into a single, final solution which fits the specification.

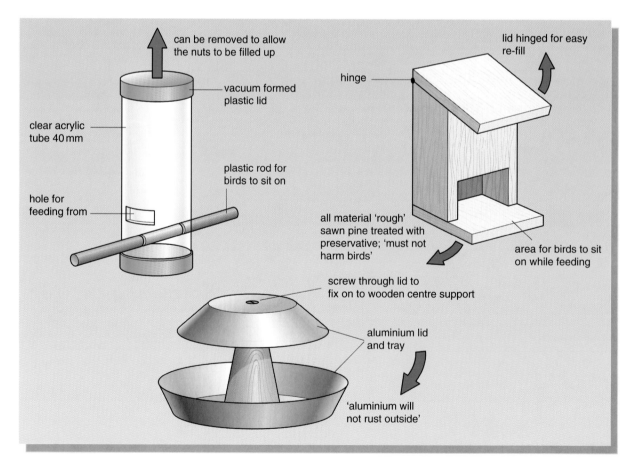

Design for a bird feeder

At this stage, it is sometimes necessary to make changes to the overall design. This may be due to material constraints such as availability or cost, or problems with the tools and equipment needed. These unforeseen problems are a valid part of the design process and you should record all of the problems you meet.

Marks awarded: 12

Developing the final design proposal involves producing design sketches and annotation with more detail. You may have to undertake further research so that the final solution meets the exact requirements of the specification. This research may be in the area of materials, joining techniques or looking at the overall sizes and dimensions.

Towards the end of the development stage you will need to assess the fitness for purpose of the final design proposal.

You should look again at the initial specification at this stage of the project. The final design proposal should be judged against the specification but with regard to its fitness-for-purpose. You should consider some of the following:

- Is it possible to visually improve the product by use of colour or texture?
- Does the shape and overall design suit the environment in which it will be used?
- Is the product fit for the purpose it was designed for?
- Can any functions be further developed?

It is also important to make sure that the materials you have chosen match up to the specification. You may have considered a number of materials and these should be evaluated against the brief and specification for the final design.

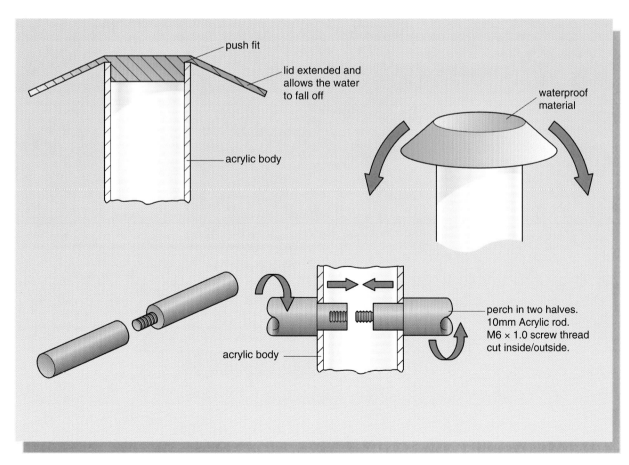

Developing the bird feeder design

Marks awarded: 12

Here are some guidelines to help you make informed decisions about the choice of materials:

- Construction – is it suitable for the construction methods suggested?
- Function – will it soak up moisture?
- Toughness – will it stand up to knocks and bumps?
- Strength – is it strong enough?
- Durability – will it last?
- Cost – is it expensive?
- Appearance – can it be finished, what colours?

Modelling and reviewing
Modelling your ideas

Modelling of ideas is a very important stage of the design process and is used to test ideas. Sometimes it is not necessary to model every aspect of your design, so a choice needs to be made about what is to be modelled.

It is important that before you do any modelling, you have a clear idea about what aspect of your product you are going to model. You should aim to test for a particular aspect of your product such as size, **aesthetics**, construction or stability.

Various techniques are used to **mock up** designs and you should consider what materials are best suited to what you are modelling. If you want to test strength or parts of the construction, then you should use identical materials that will allow you to carry out a fair test. If, however, you are modelling a 3D product which may be difficult to draw, then plasticine or modelling foam would be an appropriate material. The table opposite gives some examples of different projects and how they have been modelled.

It is important that when you undertake any modelling, you record your ideas with photographs and evaluate them, recording your findings and observations.

Reviewing your ideas

When you have completed your design ideas, you will need to evaluate them. You will also need to review your final design proposal against the initial specification. It is essential that you show evidence of having reviewed the design decisions that you have made. This can take the form of evaluative comments made by you and ideally should include the views and comments of others such as clients, experts and potential users.

Any review and design evaluation should be carried out against the specification. This will ensure that all aspects of the design are covered and evaluated.

One accurate way of presenting and recording your results is to produce a chart that gives each aspect of your design a numerical value. This method of carrying out an evaluation is known as an **attribute analysis**. If you ask intended users what they think of your designs and record their comments, this is known as a **user evaluation**.

For your project

Model one or more aspects of your final design proposal. Photograph your model and include it in your folder along with your comments.

When you have carried out your evaluation and review, you should record which design idea you have chosen and state the reasons why. You should also record any features or aspects which you may need to improve and develop further.

For your project

Evaluate your initial ideas and final design proposal against your specification. Try to get the views and comments of users too.

Marks awarded: 3

Material	Boxes	Frameworks	Handles/taps	Organic shapes
Balsa wood	✓	✓	✓	✓
Card	✓	✓		
Clay			✓	✓
Corrugated card	✓	✓		
Foam			✓	✓
MDF			✓	✓
Paper	✓			
Pipe cleaners		✓		
Wire		✓	✓	

Material	Templates	Tables	Storage units	Forging work
Balsa wood	✓	✓	✓	✓
Card	✓	✓	✓	
Clay				
Corrugated card			✓	
Foam				
MDF				
Paper			✓	
Pipe cleaners		✓		✓
Wire		✓		✓

Choosing the right modelling materials

Criteria 3

Aim

- To use written and graphical techniques including ICT and CAD (where appropriate) to generate, develop, model and communicate.

The marks in this section will be awarded for a range of graphical techniques including ICT and **CAD** where appropriate, to generate, develop, model and communicate.

Written communication

Written work, such as the design brief, specification and evaluation, will form the basis of your folder. Your writing should be clearly laid out and presented in a logical way.

You should aim to use appropriate terms when writing your coursework project. What you write reflects your knowledge and understanding. Your written work should also include the use of specialist terminology.

When **annotating** your design work, you should use specialist terminology and technical vocabulary. For example, rather than simply saying 'wood', you should

Use technical and specialist terminology to give information about:

- materials
- finishes
- components
- manufacturing techniques/processes.

state what type of wood, or the term 'painted' should be expanded to include what type of paint was used including primer, undercoat and top coat.

To be successful you will:

- Produce high quality written work throughout your project folder.
- Use specialist terminology, presented in a well thought out and logical way.

Marks awarded: 3

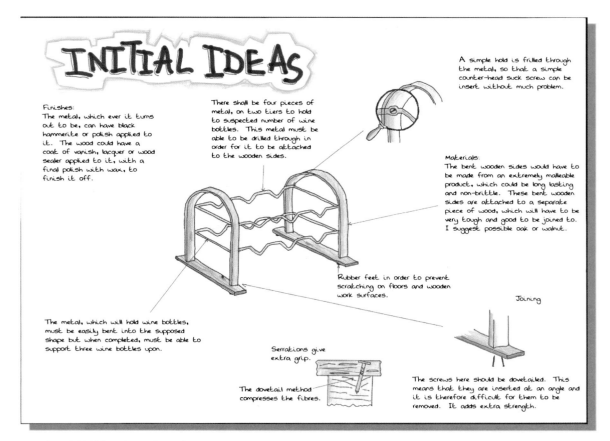

A students initial ideas for a wine rack

Information and communications technology (ICT)

There are many opportunities throughout your coursework project to use ICT and you should make good use of the resources and facilities that you have available in your school or college.

ICT is another way in which to communicate and its use should add to the content or presentation of your work. ICT can be used in your project in any of the following ways where appropriate.

Researching information

- Use a database, for example, the Internet or **CD-ROM** to seek out information.
- Use e-mail to communicate with outside agencies when seeking information.
- Present and analyse information using charts generated in a database or spreadsheet.
- Use grammar and spell checkers to correct written coursework.

Generating ideas

- Use a digital camera or scanned images to produce source material.
- Use a draw or paint software package to generate, edit and communicate design ideas.

Developing ideas

Use a 3D modelling program to produce a visual image of the proposed product.

Considering industrial applications

Use a digital camera to record the sequence of making your project.

Making

- Use a **cutter/plotter** to produce shapes in thin materials for product decoration.
- Use **CNC** machinery where appropriate, such as a milling machine, to make a number of identical components.

> **To be successful you will:**
> - Use ICT and CAD/CAM where appropriate.

Marks awarded: 3

Use a digital camera to produce source material

Other media

Graphical communication exists in many forms, several of which you will be using throughout your coursework project. Some examples include:

- freehand sketching
- formal technical drawings
- pictorial drawing
    - oblique
    - perspective
    - isometric
    - exploded views
- colour rendering
    - crayons
    - pencils
    - markers
    - water colours
- model making
    - photography.

Once you have developed and modelled your chosen idea, you will need to produce a set of drawings which will allow a client or manufacturer to see what the product looks like. Your final drawing or presentation can be produced using any of the techniques illustrated opposite.

Isometric

Gives a 3D image and looks quite realistic. It is constructed using a 30-degree set square, but it is difficult to draw curves.

Oblique

Based on a 45-degree set square. Curves are easier, but it often looks distorted.

Perspective

The most realistic of all 3D images, it can be viewed from above or below. It is quite difficult to draw curves.

Orthographic

Has three separate viewing positions, allowing complicated products to be drawn quite easily. This type of drawing is easily dimensioned, but it can sometimes be difficult to visualize the final product.

Exploded drawings

Used to show construction and assembly details. Parts are easily identified along with any components used.

Working drawings

Working drawings are used to show someone how to make your product. They provide lots of information about dimensions and parts. These types of drawings are now widely produced on a **CAD** system. This has many benefits for those in industry and school:

- The drawings are saved on disk for easy access and retrieval.
- Changes can be made easily.
- The drawings can be emailed anywhere in the world.
- The system can often be used to generate 3D pictorial drawings.
- Some CAD programs will output to CNC equipment such as lathes and milling machines.

■ **Hints and tips** ■

Good quality communication skills are essential in order to explain your ideas and concepts to others.

For your project

When you are ready to complete your final presentation drawing, use a pencil and instruments. Make a few photocopies of it and try rendering it using a variety of colouring techniques such as crayons, markers, pastels or water colours. Cut around the best one and stick it into your folder.

To be successful you will:

- Use a range of graphical media and techniques in your work.

Marks awarded: 3

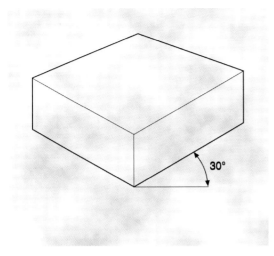

Isometric

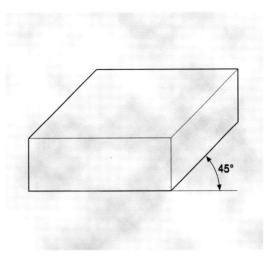

Oblique

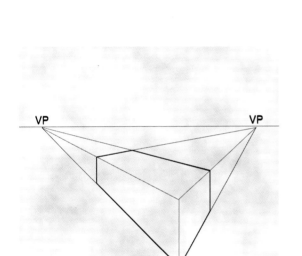

Perspective

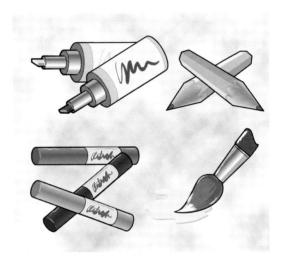

Drawing tools

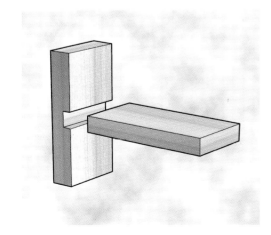

Exploded drawing of a wood joint

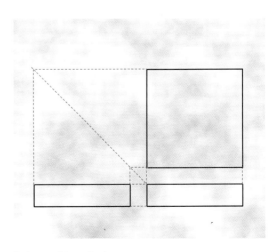

Orthographic projection

Criteria 4

Aims

- To produce and use a detailed working schedule which includes a range of industrial applications as well as the concepts of systems and control.
- To demonstrate an understanding of industrial processes and use them in your work where appropriate.

Systems and control

In order to make sure that your project is successful, it is essential that you apply some of these systems and control measures to your work. They will also help you to plan, organize and ensure that things are completed in the correct order.

The systems approach to managing and controlling the manufacture of your project involves four main areas:

- inputs
- process
- outputs
- feedback.

A flowchart is one way of planning the manufacture of your project.

At critical stages in the manufacture of your project, you will need to carry out some checks for quality and accuracy of manufacture. To do this, you will need to use a decision box (see below).

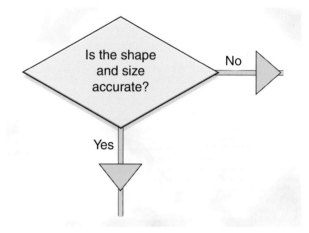

A decision box

The checks should be made against some sort of reference or dimension. For example, if you need to make a component or piece to a known dimension, you will need to measure this piece to check how accurate it is. If it is too big, you can make it a little smaller. If it is too small, you will need to make another.

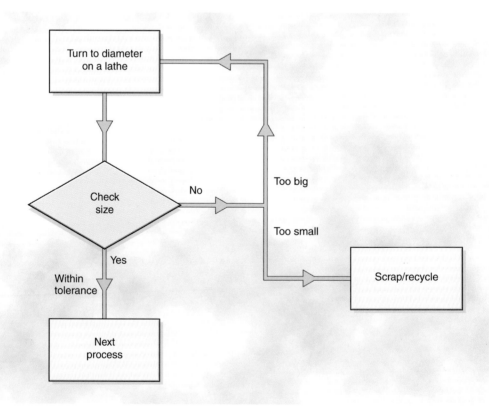

You can plan the manufacture of your project with a flowchart

Quality control check

The decision box is a **quality control** check. This type of system is called a closed loop system. It is one in which constant checks are being made to influence subsequent actions within the manufacturing process such as checking the dimensions and accuracy or the quality of a surface finish.

At this point, it is sufficient simply to label the 'process' block and to record what process or processes are being used.

The 'inputs' should indicate the raw materials selected and the components used.

The 'output' is a part or product which has been manufactured using the materials and processes.

A simple outline flowchart is shown below.

When you come to draw up a list for the manufacturing of your product, you should break down the making stages into a series of sub-systems of inputs, processes and outputs. The purpose of these systems is to change inputs into outputs through a series of processes. See the simple outline flowchart to the right for an example of this.

As you produce your outline systems diagram for your manufacture, explain the inputs, processes, outputs and feedback at each stage.

The use of feedback should be used at checking stages as illustrated in the example. Any quality control checks will need to be made against a certain standard or set of dimensions. If the item does not come up to the standard or fall within the tolerance limits, then it will either have to be scrapped or recycled.

> **To be successful you will:**
> - Outline systems diagrams for the manufacture of your project, showing inputs, processes, outputs and feedback.

Marks awarded: 3

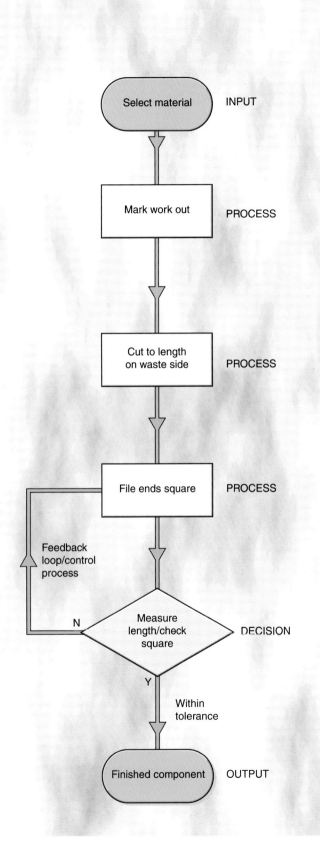

A simple outline of a flowchart

Schedules
Schedules

Essential to the whole process of manufacture is careful planning. Plans should be in enough detail to enable another person to make the final outcome from the schedule.

The basis of any schedule is a flowchart or a **Gantt chart**. See the example of a Gantt chart to the right. The chart should include details such as:

- the main stages in production
- the collecting of materials and components
- the preparation of materials to include measuring and marking out
- the manufacturing processes
- assembly details
- finishing.

You should also provide an adequate set of drawings showing:

- the individual pieces required to make the product
- the dimensions of the separate pieces
- how they fit together (an assembly drawing)
- dimensions and materials.

task		week				
		1	2	3	4	5
1	MDF sides					
2	MDF curves					
3	sand and finish					
4	assembly					
	time allowed (2.5 hours per week of lessons)	2.5	2.5	2.5	2.5	2.5
					total time	12.5

A simple Gantt chart

The various types of drawing that this involves will be given credit in Criteria 3 (see pages 126–9).

> **To be successful you will:**
> - Produce and use a detailed working schedule.

Marks awarded: 3

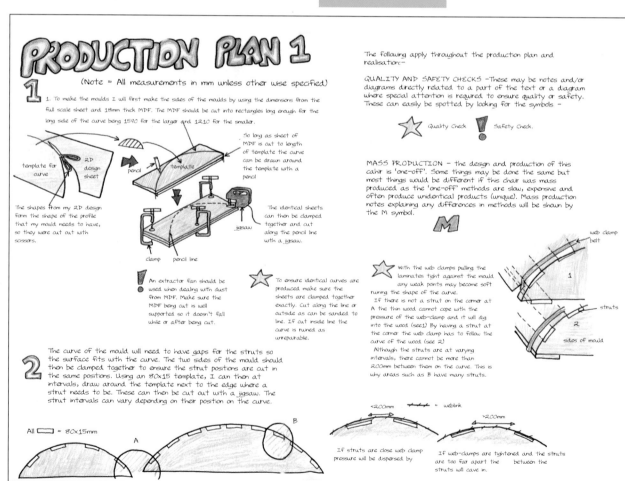

A student production plan

Full course coursework

The schedule should be viewed as an aid to working, so it must indicate the order in which parts and sub-systems need to be made and it should provide as much detail as possible. Details such as any **jigs**, moulds or **templates** used, or any **CAD/CAM** employed to produce identical parts or sub-systems should also be included. Other considerations could include scale of production, timescales and quality issues.

Industrial applications

The manufacture of a single product may involve a number of different manufacturing techniques and processes. It may also involve a range of different materials and components. The bringing together of all these different pieces must be carefully planned and organized to ensure that everything is assembled in the correct sequence. This procedure is carefully controlled and monitored within industry and is part of the company's overall quality assurance and control procedures.

Having produced a one-off product, you should consider the demands on equipment and processes of using **batch production** to produce a few hundred of the same product.

You should try to include some detail about:
- the materials that you would use and any other components
- reasons for the choice of materials
- reasons concerning the choice of finishes.

You also need to refer to the processes that could be used. This may be in the form of notes, diagrams or pictures.

To be successful you will:
- Demonstrate in your folder an understanding of a range of industrial applications and processes.

Marks awarded: 3

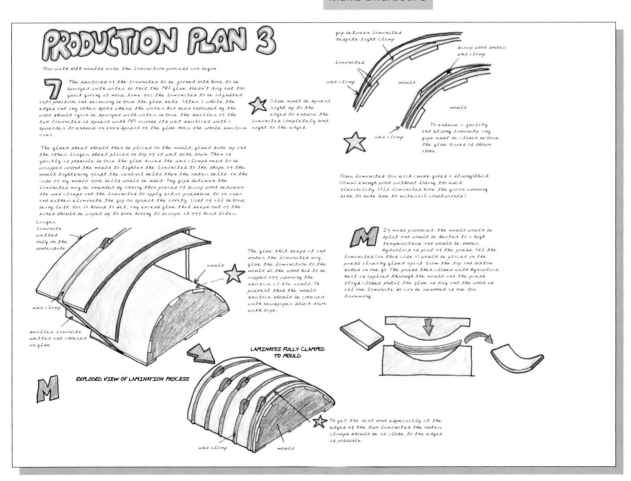

Student work showing more detail of components

Criteria 5

Aims

- To select and use tools, equipment and processes effectively and safely to make single products and products in quantity.
- To use CAM appropriately.

▪ Hints and tips ▪

This specific area of selecting and using tools, making products and working safely is worth more marks than any of the other sections.

You should therefore spend an appropriate amount of your time making sure that you complete this section to the best of your ability.

Select and use tools, equipment and processes

Once you have planned the manufacture of your product, you should be ready to turn your idea into reality (also known as **realization**).

You may well find this the most enjoyable and most exciting part of Design and Technology. If you have planned and prepared thoroughly, then the realization of your project should be straightforward.

The stages of manufacture will have been identified and broken down in the detailed working schedule. Each job such as marking out, cutting and joining, has its own special tools.

You will need to consider carefully which tools to choose. This will lead to greater accuracy when starting to cut and join separate parts together.

▪ Hints and tips ▪

If you are carrying out a particular technique or new process that you have not done before, practise on a scrap piece of material before carrying it out on your project.

There may be times when you need to modify (alter) your idea due to limitations and constraints placed on you. This might be as a result of the materials or the manufacturing processes you have available to you in your school workshop. Record any modifications you make due to these constraints in your project folder. Include photographic evidence of using tools and equipment. This could be a photographic storyboard. Give information on why the tools and materials were selected.

To be successful you will:

- Select and use tools and equipment as expertly as possible.
- Make modifications to techniques and processes where necessary.

Marks awarded: 18

Make products

When realizing your coursework project, you need to consider the following key areas:

- the production of a high quality product
- a fully functional product which meets the requirements of the specification
- the exploration of unfamiliar construction techniques if appropriate
- the use of **CAM** if available and relevant
- safety awareness for yourself and others.

Quality is a key issue throughout the project. In both your folder and the practical realization you should always aim to produce work of the highest quality. This starts with the selection of materials and the initial marking out stages.

Computer-aided manufacture (CAM)

All the manufacturing discussed so far has been related to manual processing and the handling of tools and materials. The automated making operations are known as computer aided manufacture (**CAM**). **CNC** lathes and milling machines are examples of these.

CAM can be used to make single items such as patterns or moulds for casting or **vacuum forming** on a CNC mill. A CNC lathe might be used to make the turned components for a G-clamp such as the foot and the threaded bar.

If you have any CAM facilities in your school and part of your project can be made using either a CNC milling machine or lathe, then you should do so. However, the use of CAM should be relevant and should not be included for its own sake.

Making gives you an opportunity to demonstrate skills in manipulating tools and equipment. High quality projects demand high level skills and you should explore and experiment with new and unfamiliar construction techniques and processes if appropriate.

■ Hints and tips ■

Practical work that is challenging and of a good quality will be rewarded more highly than good quality work that is more straightforward.

▌ To be successful you will:

- Produce a high quality product that is fully functional and meets the requirements of the specification.
- Use CAM appropriately if relevant.
- Produce good quality photographs of the final product.

Marks awarded: 18

Work Safely

Safety is also very important when working with any tools, materials and machinery. You should take into account the handling of tools and any dangerous or hot materials. Protective equipment and clothing should always be worn. Any safety precautions relating to specific machines must be followed.

Safety awareness should be identified and recorded as part of the schedule for the planning of manufacture. More detailed risk assessments and analysis of hazards should be included in your folder work. For example: this is what one student wrote:

Risk assessment

When it came to the casting of my project, I had to make sure that I followed all the safety procedures. The tools had to be warmed before use. The surrounding area was clear and the floor space tidy. All safety equipment had to be worn like the face shield, leather gloves, apron and shoe covers.

Although casting can be quite dangerous, attention to detail, like in the example above, should ensure that accidents do not occur.

Other safety issues have to be observed in the workshop such as the handling and carrying of sharp tools. You should record this in the schedule or in the production details.

Regardless of how confident you may be, or if you are in a rush, you must always obey any safety instructions given by your teacher. You must also observe all safety signs.

For your project

Highlight and record evidence of your safety awareness.

Include details of risk assessment and any analysis of hazards.

▌ To be successful you will:

- Demonstrate a regard for safety awareness for yourself and others when working with materials, tools and equipment.

Marks awarded: 3

Criteria 6

Aims

- To devise and apply tests to check the quality of work at critical control points.
- To evaluate the product to ensure it is of suitable quality for the intended use.
- To suggest modifications to improve the product's performance.

■ Hints and tips ■

Testing and evaluation is generally the most poorly completed section in projects. This is because not enough time is left at the end of the project to carry out any meaningful testing, evaluation and writing up. Make sure you allow sufficient time in your project for this.

Tests and checks

Testing and checking should be in evidence throughout your coursework project folder. You will probably have carried out some very basic tests on any models you made in the development section such as stability, construction methods or key dimensions and sizes. The findings from these initial tests should have been recorded at the time and used to further develop or make clear your design thinking. The final tests to be undertaken will be carried out when the product has been completed.

The testing procedures need to be thoroughly planned, carried out and your results and findings recorded. They can then be used to suggest further developments or modifications. The basis for much of the testing should be against all aspects of the specification and the user's needs. The tests should be conducted in a controlled manner and in the position or environment in which the product will normally be used.

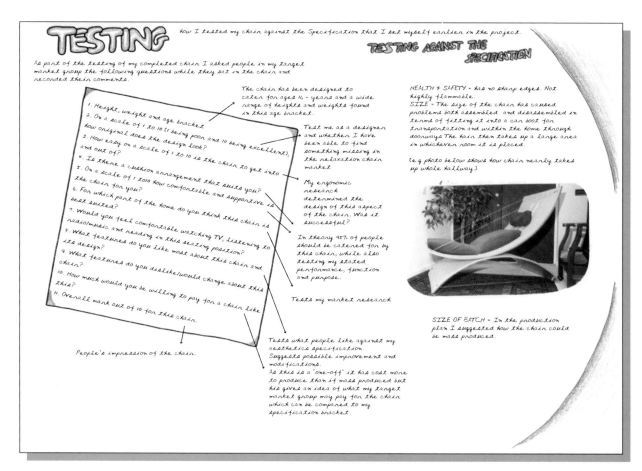

A student tests against the specification

These are some general questions that could be asked of almost every product:

- Does it fit in with its surroundings?
- Is it easy to move or adjust?
- Is it too big or too small?
- Is it safe to use?
- Is it clear to the user how to use the product?

There will be more specific questions related to your product, but you should leave no aspect overlooked.

You should also check your final product against the initial specification. For each of the individual specification points, you should comment upon how successful your solution has been. It is also useful to evaluate the specification in terms of how thorough and adequate it was in the first place.

User trials or field tests are another way of conducting tests. You can ask a likely user for his or her views and record the person's comments on whether he or she liked it or not and how it might be improved.

> ### I To be successful you will:
> - Devise and apply tests to check quality of your work at critical points as well as at the end.
> - Ensure that your produts are of a suitable quality for the intended use.

Marks awarded: 3

> ### ■ Hints and tips ■
> There are a number of key areas of tests and checks that you need to make and you should use the following as a simple checklist:
>
> - test product against specification
> - measure its fitness-for-purpose
> - field testing under working conditions
> - testing over extended periods
> - third party user testing
> - testing against external standards.

Evaluate product

The final product evaluation should take the form of a written report. The product should be objectively evaluated against the initial design specification. This will help to justify the success of your project. The whole project also needs to be evaluated against:

- how you managed your time
- any problems encountered
- how the problems were overcome
- modifications made in the manufacturing stages.

All testing (such as user trials) and evaluation provides you with feedback on the product's performance and fitness-for-purpose. This data and information can be collected and presented in a number of ways:

- questionnaires
- interviews
- written comments
- test data.

When this information is collected and presented, it should be used to suggest design improvements and modifications in the light of your experience. Details concerning modifications for future development should be taken from this exercise and recorded formally in your coursework folder. Suggestions for improvement could focus around a number of these areas:

- product performance
- quality of manufacture and design
- fitness for purpose
- target market
- larger scale production.

There are a number of general questions you could ask when evaluating your product:

- Did the design brief allow you enough scope?
- Was the initial specification too narrow (restrictive) or too wide?
- Did the product cost more than initially expected?
- Does it work well?
- Could it be made cheaply in large quantities?
- Is it of a good enough quality?
- What is its life expectancy?
- Did the project run to plan in terms of time?
- What would you do differently if you could do the project again?
- Do you feel you put in enough effort?

> ### I To be successful you will:
> - Evaluate your product not just in your own opinion but using your test results and the views of the user.

Marks awarded: 3

Modifications

The process of testing, checking and evaluating your project provides valuable feedback on performance and fitness for purpose. When you have completed all of these aspects you should be able to suggest design improvements and modifications for future development of the product. It may also be possible to suggest improvements to the manufacturing processes involved. Quite often the two are interrelated in that a change to the design will have an impact on the manufacturing processes.

Suggestions for improvement should relate to all of the following areas:

- product performance
- quality of design
- fitness-for-purpose
- target market
- larger scale production.

The key to a successful project is management of time and resources. Testing, evaluating and suggesting modifications cannot be carried out effectively on incomplete work.

> **To be successful you will:**
> - Suggest modifications that would improve the performance of your product.

Marks awarded: 3

■ **Hints and tips** ■

- Discuss your project with your teacher to make sure that it is appropriate to your ability in the time available.
- Don't be over ambitious.
- Don't work on joint projects unless you can justify individual marks in all areas of assessment.
- Do look at how much each separate assessment criteria is worth
  – target the marks in each of the areas and divide your time appropriately
  – make sure you spend enough time on the two largest sections.
- Make use of clear and effective photographs throughout the modelling stages and of the practical outcomes.
- Plan your work making sure you know what has to be completed by when – your teacher will confirm when the completed coursework has to be handed in.
- Stick to the recommended number of pages.

Finally, make sure you enjoy the experience!

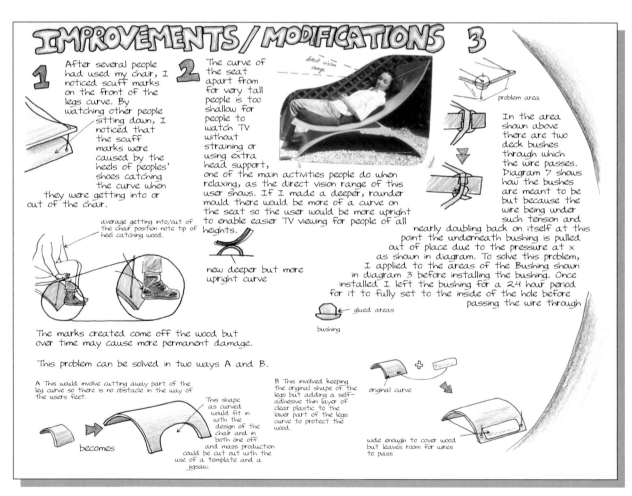

A student looks at ways of modifying and improving a chair design.

Short course coursework

Criteria 1

Aim

- To identify needs and to use information sources to develop specification criteria.

> ### ■ Hints and tips ■
>
> The information you use must be relevant and targeted specifically to your project. Useless padding does not gain marks!

> ### ■ Hints and tips ■
>
> Take some time to consider carefully all the projects that you have come up with. Discuss the project titles with your teacher in order to make sure that the project will allow you to complete all the assessment criteria.

In industry, a lot of time and effort is spent on product research and development. Many important questions are asked such as:

- Is there a real need for the product?
- Who are the potential users?
- What is the target market group?
- Is the market large enough?
- What environment will the product be used in?
- What existing products are already available?
- What are the opinions of the users of a good and a bad product?

It is certainly worth starting with some of these points in the initial stages of your research. It might also be useful to interview somebody about your product.

Identifying a need

Designers design new products for a number of reasons:

- to solve a problem
- to improve the performance of an existing product which may function poorly, be unreliable or simply look out of date
- to redesign a product because a new technology has been introduced
- to improve the sales of an existing product by improved graphics or new colours and packaging.

Any of the above reasons can help you to decide upon a situation for your design project. Brainstorm some ideas for your own situation. Start with areas such as play, schools, leisure or the home. Then add in products to do with these situations, or any other areas that you may be interested in.

Rather than brainstorming your own ideas, there is also a bank of titles available that can be used as starting points for your project. Your teacher will have a copy of these.

Producing a design brief

Once you have gathered and analysed all your information, you will need to produce a design brief. This is a short statement about what you are intending to design and make. It states the problem and provides some detail about who it affects, what happens as a result, when it occurs and why it is a problem.

In analysing and summarizing the information you have gathered you must justify that the market group has a particular need(s) and produce a design brief.

Having produced a design brief, you should now carry out any further research and analysis. This will enable you to produce a product specification, which will guide you through the design process and product development.

For your project

Using the various sources of information you have available, gather useful, relevant information on your chosen project.

Analyse your data and select the most relevant and appropriate material and present it on page 2 of your project folder as 'A Summary of Information Gathered'.

Gathering information

Before starting to design and make your product, it is important to do your research thoroughly. Gather information from a wide range of sources to give you as much useful data as possible. Possible sources you could use might be:

- market research
- consumer surveys
- visits to manufacturers
- product test reports and magazines
- databases
- the Internet
- data sheets.

Producing a specification

A product specification is a list of the product's main functions and qualities. The specification should contain quantitative and qualitative information.

Quantitative information can be measured in many ways and would include details such as a maximum weight or the overall dimensions.

Qualitative information provides details such as:

- the materials
- the purpose of the product
- scale of production
- appearance
- safety factors
- product maintenance
- environmental issues.

Your product specification needs to include enough detail to guide you in your thinking and be a basis for generating ideas. It may be that as you start designing, your specification changes. You should keep all versions of the specification to show the development and refinement of the design.

The specification should also be used when you evaluate your initial ideas.

For your project

Present a specification that describes the form, function, user requirements and budgetary constraints for your design.

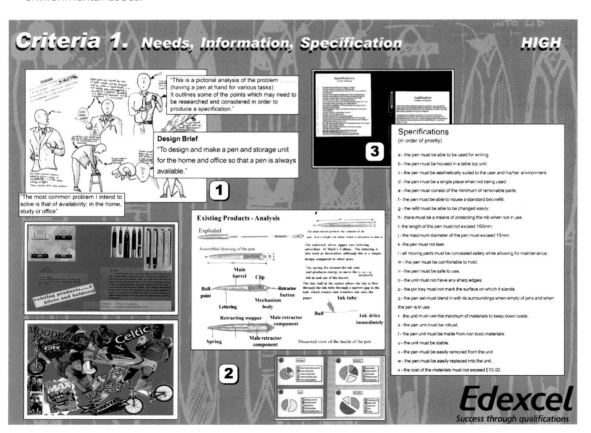

Criteria 2

Aims

- To develop ideas from the specification.
- To check, review and modify ideas as necessary to develop a product.
- To develop and model ideas to produce a real-istic design proposal.

Ideas

During this stage, you are expected to generate a wide range of feasible design ideas. Your ideas should be based on the points raised in your specification.

It is very important that your design ideas are clearly presented so that they can be easily understood. They should also show your ability to think creatively. Notes need to be added to your ideas to give details about the materials used, how things work and what finishes are to be used. Writing notes about your work is called **annotation**.

For help with ideas, try looking at:

- natural forms – such as leaves, shells and fossils
- the work of other designers – such as Starck and Alessi
- a period of history or a design movement – such as Arts and Crafts or Bauhaus
- a type of music or a fashion
- a theme, for example a type of music or a fashion
- new technologies – such as new materials like 'polymorph' or a **shape memory alloy (SMA)**.

You may also look at existing products, but make sure that you show some development and do not simply copy them.

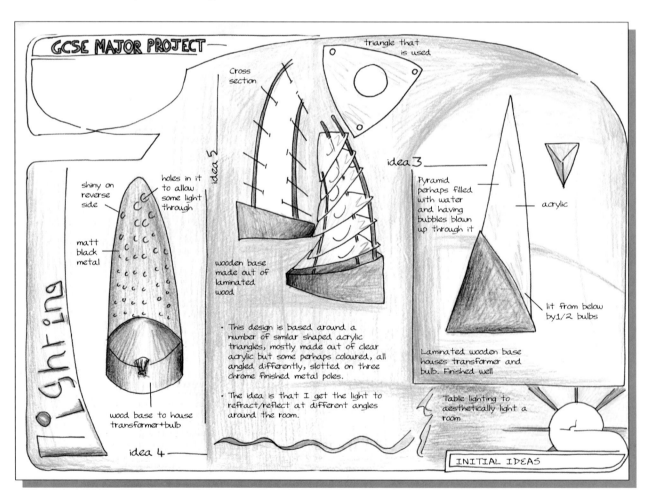

A student's initial ideas with annotations

Number of pages: 1

Marks awarded: 12

Opposite is an example of some annotated ideas.

Development of ideas

Idea development brings together the best features from the initial ideas into a single, final solution which fits the specification.

At this stage, it is sometimes necessary to make changes to the overall design. This may be due to material constraints such as availability or cost, or problems with the tools and equipment needed. These unforeseen problems are a valid part of the design process and you should record all of the problems you meet.

Other factors to consider are:

- construction
- cost
- appearance
- function.

Modelling your ideas

Modelling of ideas is a very important stage of the design process and is used to test ideas. Sometimes it is not necessary to model every aspect of your design, so a choice needs to be made about what is to be modelled.

Various techniques are used to **mock up** designs and you should consider what materials are best suited to what you are modelling. If you want to test strength or parts of the construction, then you should use identical materials that will allow you to carry out a fair test.

If, however, you are modelling a 3D product which may be difficult to draw, then plasticine or modelling foam would be an appropriate material. The table on page 125 gives some examples of different projects and how they have been modelled.

For example, if you were making a CD rack, it would be a good idea to look at the overall shape and proportion, the stability and the construction methods. It is important that when you undertake any modelling, you record your ideas with photographs and evaluate them.

Number of pages: 1

Total number of pages for this whole section could be as many as 4

Marks awarded: 12

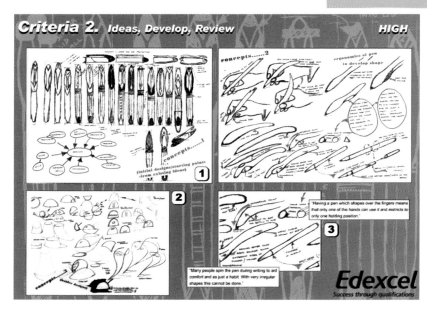

Criteria 2. *Ideas, Develop, Review*   HIGH

Edexcel
Success through qualifications

Criteria 3

Aims

The marks in this section will be awarded for a range of graphical techniques including ICT and CAD where appropriate, to generate, develop, model and communicate.

Written communication

Written work, such as the design brief, specification and evaluation, will form the basis of your folder. Your writing should be clearly laid out and presented in a logical way.

You should aim to use appropriate terms when writing your coursework project. What you write reflects your knowledge and understanding. A high quality of written work should also include use of specialist terminology.

You should highlight specific materials and give details of manufacturing techniques and processes. You should also give accurate details about the selection of finishes and the use of any component such as screws, nuts and bolts or any other type of fixing.

■ Hints and tips ■

Use technical and specialist terminology to give information about:

- materials
- finishes
- components
- manufacturing techniques/processes.

To be successful you will:

- Use technical language where appropriate.
- Be competent in the use of specialist terminology
- Present information in a well thought out and logical way.

Number of pages: evidence throughout folder

Marks awarded: 3

Information and communications technology (ICT) and other media

There are many opportunities throughout your coursework project to use ICT and you should make good use of the resources and facilities that you have available in your school or college.

ICT is another way in which to communicate and its use should add to the content or presentation of your work. ICT can be used in your project in any of the following ways where appropriate.

Researching information

- Use a database, for example, the Internet or **CD-ROM** to seek out information.
- Use e-mail to communicate with outside agencies when seeking information.
- Present and analyse information using charts generated in a database or spreadsheet.
- Use grammar and spell checkers to correct written coursework.

Generating ideas

- Use a digital camera or scanned images to produce source material.
- Use a draw or paint software package to generate, edit and communicate design ideas.

Developing ideas

Use a 3D modelling program to produce a visual image of the proposed product.

Considering industrial applications

Use a digital camera to record the sequence of making your project.

Making

- Use a **cutter/plotter** to produce shapes in thin materials for product decoration.
- Use **CNC** machinery where appropriate, such as a milling machine, to make a number of identical components.

Other media

Graphical communication exists in many forms, several of which you will be using throughout your coursework project. Some examples include:

- freehand sketching
- formal technical drawings

- pictorial drawing
  - oblique
  - perspective
  - isometric
  - exploded views
- colour rendering
  - crayons
  - pencils
  - markers
  - water colours
- model making
- photography.

Once you have developed and modelled your chosen idea you will need to produce a set of drawings which will allow a client or manufacturer to see what the product looks like. This should be presented on the sheet entitled 'The Chosen Design', which your teacher will provide you with, on which you must give some information about the form, size, construction, materials and finish.

Your final drawing or presentation can be produced using any of the techniques illustrated on page 129.

Isometric

Gives a 3D image and looks quite realistic. It is constructed using a 30-degree set square but it is difficult to draw curves.

Oblique

Based on a 45-degree set square. Curves are easier but it often looks distorted.

Perspective

The most realistic of all 3D images, it can be viewed from above or below. It is quite difficult to draw curves.

Orthographic

Has three separate viewing positions, allowing complicated products to be drawn quite easily. This type of drawing is easily dimensioned but it can sometimes be difficult to visualize the final product.

Exploded drawings

Used to show construction and assembly details. Parts are easily identified along with any components used.

Working drawings

Working drawings are used to show someone how to make your product. They provide lots of information about dimensions and parts. These types of drawings are now widely produced on a CAD system. This has many benefits for those in industry and school:

- The drawings are saved on disk for easy access and retrieval.
- Changes can be made easily.
- The drawings can be emailed anywhere in the world.
- The system can often be used to generate 3D pictorial drawings.
- Some CAD programs will output to CNC equipment such as lathes and milling machines.

For your project

When you are ready to complete your final presentation drawing, use a pencil and instruments. Make a few photocopies of it and try rendering it using a variety of colouring techniques such as crayons, markers, pastels or water colours. Cut around the best one and stick it into your folder.

To be successful you will:
- Show evidence of a variety of drawing styles.
- Use ICT where appropriate.

Number of pages: evidence throughout folder

Marks awarded: 3

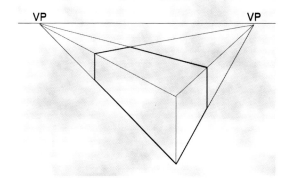

Perspective

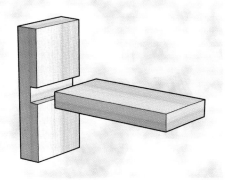

Exploded drawing of a wood joint

Aims

- To produce and use a detailed working schedule which includes a range of industrial applications as well as the concepts of systems and control.
- To demonstrate an understanding of industrial processes and use them in your work where appropriate.

Systems and control

The manufacture of a single product may involve a number of different manufacturing techniques and processes. It may also involve a range of different materials and components. The bringing together of all these different pieces must be carefully planned and organized to ensure that everything is assembled in the correct sequence. This procedure is carefully controlled and monitored within industry and is part of the company's overall quality assurance and control procedures.

In order to make sure that your project is successful, it is essential that you apply some of these systems and control measures to your work. They will also help you to plan, organize and ensure that things are completed in the correct order.

The systems approach to managing and controlling the manufacture of your project involves four main areas:

- inputs
- process
- outputs
- feedback.

A flow chart is one way of planning the manufacture of your project.

At critical stages in the manufacture of your project, you will need to carry out some checks for quality and accuracy of manufacture. To do this, you will need to use a decision box (see below). The checks should be made against some sort of reference or dimension. For example, if you need to make a component or piece to a known dimension, you will need to measure this piece to check how accurate it is. If it is too big, you can make it a little smaller. If it is too small, you will need to make another.

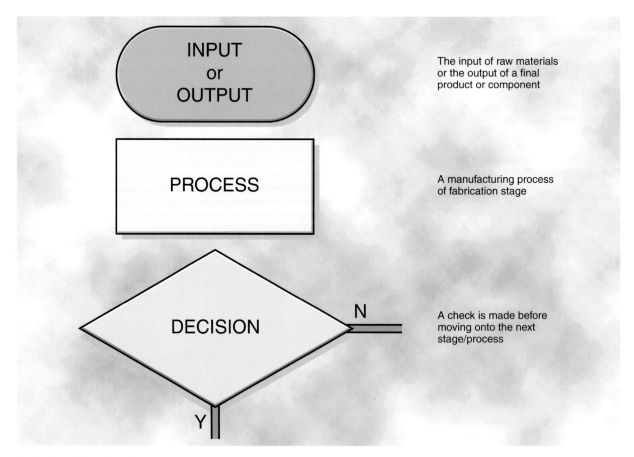

Symbols used in a flow chart

The decision box is a quality control check. This type of system is called a closed loop system. It is one in which constant checks are being made to influence subsequent actions within the manufacturing process such as checking the dimensions and accuracy or the quality of a surface finish.

At this point, it is sufficient simply to label the 'process' block and to record the process or processes being used.

The 'inputs' should indicate the raw materials selected and the components used.

The 'output' is a part or product which has been manufactured using the materials and processes.

> **To be successful you will:**
> - Outline systems diagrams for the manufacture of your project, showing inputs, processes, outputs and feedback.
> - Include a work schedule.

Marks awarded: 3

A simple outline flow chart is shown to the right.

Industrial application

The manufacture of a single product may involve a number of different manufacturing techniques and processes. It may also involve a range of different materials and components. The bringing together of all these different pieces must be carefully planned and organized to ensure that everything is assembled in the correct sequence. This procedure is carefully controlled and monitored within industry and is part of the company's overall quality assurance and control procedures. Having produced a one-off product, you should consider the demands on equipment and processes of using batch production to produce a few hundred of the same product.

> **To be successful you will:**
> - Demonstrate in your folder an understanding of a range of industrial applications and processes.

Number of pages: 1 in total for systems control and industrial applications

Marks awarded: 3

You should try to include some detail about:
- the materials that you would use and any other components
- reasons for the choice of materials
- reasons concerning the choice of finishes.

You also need to refer to the processes that could be used. This may be in the form of notes, diagrams or pictures.

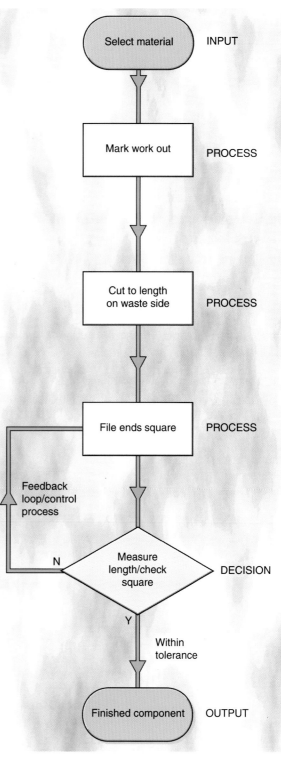

A simple outline of a flow chart

Criteria 5

Aims

- To select and use tools, equipment and processes effectively and safely to make single products and products in quantity.
- To use CAM appropriately.

■ Hints and tips ■

This specific area of selecting and using tools, making products and working safely is worth more marks than any of the other sections. You should therefore spend an appropriate amount of your time making sure that you complete this section to the best of your ability.

Select and use tools, equipment and processes

Once you have planned the manufacture of your product, you should be ready to turn your idea into reality (also known as **realization**). Hopefully, you will find this the most enjoyable and most exciting part of Design and Technology. If you have planned and prepared thoroughly, then the realization of your project should be straightforward.

The stages of manufacture will have been identified and broken down in the detailed working schedule. Each job such as marking out, cutting and joining, has its own special tools.

You will need to consider carefully which tools to choose. This will lead to greater accuracy when starting to cut and join separate parts together.

■ Hints and tips ■

If you are carrying out a particular technique or new process that you have not done before, practise on a scrap piece of material before carrying it out on your project.

There may be times when you need to modify (alter) your idea due to limitations and constraints placed on you. This might be as a result of the materials or the manufacturing processes you have available to you in your school or college workshop. Record any modifications that you make due to these constraints in your project folder.

To be successful you will:

- Select and use tools and equipment as expertly as possible.
- Make modifications to techniques and processes where necessary.
- Show photographs of tools and equipment being used.

Number of pages: 1

Marks awarded: 18

Make products

When realizing your coursework project, you need to consider the following key areas:

- the production of a high quality product
- a fully functional product which meets the requirements of the specification
- the exploration of unfamiliar construction techniques if appropriate
- the use of **CAM** if available and relevant
- safety awareness for yourself and others.

Quality is a key issue throughout the project. In both your the folder and the practical realization you should always aim to produce work of the highest quality. This starts with the selection of materials and the initial marking out stages.

Computer-aided manufacture (CAM)

All the manufacturing discussed so far has been related to manual processing and the handling of tools and materials. The automated making operations are commonly known as computer aided manufacture (**CAM**). **CNC** lathes and milling machines are examples of these.

CAM can be used to make single items such as patterns or moulds for casting or vacuum forming on a CNC mill. A CNC lathe might be used to make the turned components for a G-clamp such as the foot and the threaded bar.

If you have any CAM facilities in your school and part of your project can be made using either a CNC milling machine or lathe, then you should do so. However, the use of CAM should be relevant and should not be included for its own sake.

> ### To be successful you will:
>
> - Produce a high quality product that is fully functional and meets the requirements of the specification.
> - Use CAM appropriately if relevant.

Number of pages: 1

Marks awarded: 18

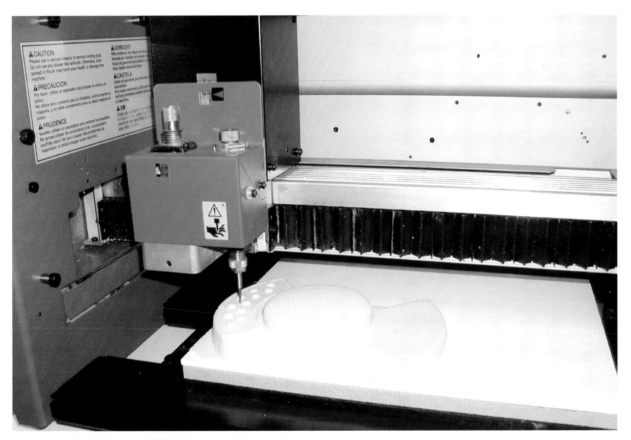

A 3D product being cut on a CAMM2 for use as a vacuum form mould

Criteria 6

Aims

- To devise and apply tests to check the quality of your work at critical control points.
- To evaluate the product to ensure it is of suitable quality for the intended use.
- To suggest modifications that would improve the product's performance.

Tests and checks

Testing and checking should be in evidence throughout your coursework project folder. You will probably have carried out some very basic tests on any models you made in the development section such as stability, construction methods or key dimensions and sizes. The findings from these initial tests should have been recorded at the time and used to further develop or make clear your design thinking.

The final tests to be undertaken will be carried out when the product has been completed.

■ Hints and tips ■

Testing and evaluation is generally the most poorly completed section in projects. This is because not enough time is left at the end of the project to carry out any meaningful testing, evaluation and writing up. Make sure you allow sufficient time in your project for this.

The testing procedures need to be thoroughly planned, carried out and your results and findings recorded. They can then be used to suggest further developments or modifications.

The basis for much of the testing should be against all aspects of the specification and the user's needs. The tests should be conducted in a controlled manner and in the position or environment in which the product will normally be used.

These are some general questions that could be asked of almost every product:

- Does it fit in with its surroundings?
- Is it easy to move or adjust?
- Is it too big or too small?
- Is it safe to use?
- Is it clear to the user how to use the product?

There will be more specific questions related to your product but you should leave no aspect overlooked.

You should also check your final product against the initial specification. For each of the individual specification points, you should comment upon how successful your solution has been. It is also useful to evaluate the specification in terms of how thorough and adequate it was in the first place.

User trials or field tests are another way of conducting tests. You can ask a likely user for his or her views and record the person's comments on whether he or she liked it or not and how it might be improved.

To be successful you will:

- Devise and apply tests to check the quality of your work at critical points as well as at the end.
- Ensure that your products are of a suitable quality for the intended use.

Number of pages: 1

Marks awarded: 3

Evaluation

The final product evaluation should take the form of a written report. The product should be objectively evaluated against the initial design specification. This will help to justify the success of your project. The whole project also needs to be evaluated against:

- how you managed your time
- any problems encountered
- how the problems were overcome
- any modifications made in the manufacturing stages.

All testing (such as user trials) and evaluation provides you with feedback about the product's performance and its fitness-for-purpose. This data and information can be collected and presented in a number of ways such as:

- questionnaires
- interviews
- written comments
- test data.

However this information is collected and presented it should be used to suggest design improvements and modifications in the light of your experience. Some details concerning modifications for future development should be drawn from this exercise and recorded formally in your coursework folder.

Suggestions for improvement could focus around a number of these areas:

- product performance
- quality of manufacture and design
- fitness-for-purpose
- target market
- larger scale production.

There are a number of general questions you could ask when evaluating a product:

- Did the design brief allow you enough scope?
- Was the initial specification too narrow (restrictive) or too wide?

- Did the product cost more than initially expected?
- Does it work well?
- Could it be made cheaply in large quantities?
- Is it of a good enough quality?
- What is its life expectancy?
- Did the project run to plan in terms of time?
- What would you do differently if you could do the project again?
- Do you feel you put in enough effort?

Number of pages: 1

Marks awarded: 3

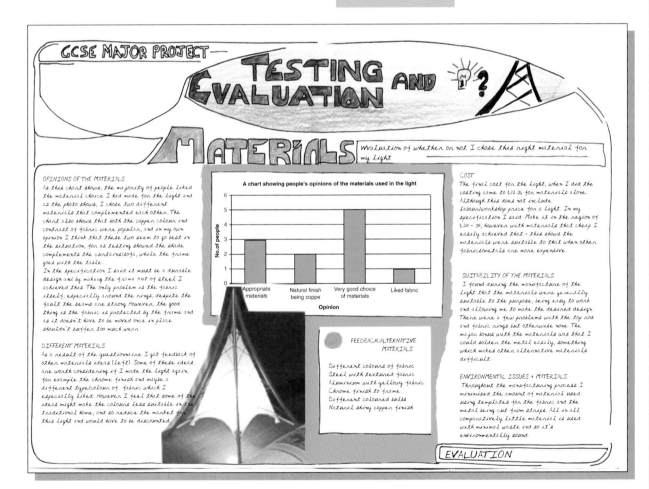

A student tests and evaluates materials for a light designed like the Eiffel Tower

Glossary

aesthetics how we respond to the visual appearance of a product, in relation to its form, texture, smell and colour

alloys metals made by combining two or more metallic elements; usually to make stronger or more resistant

alumina the material which is formed as a result of refining bauxite (aluminium ore)

annotation explanatory notes on a design, detailing such things as the materials used, how it works

anthropometrics the study of the human form in relation to size, movement and strength; used in ergonomics

attribute analysis a method of evaluating the key characteristics and features of a design/product by giving each aspect of the design a numerical value

batch production a method of production where a number of components are made all at once. Repeated batches are sometimes made over a longer period of time

billets bars of metal that need further processing

black mild steel steel which has been rolled hot and therefore the surface is left black and slightly rough

blast furnace a vessel in which iron ore, lime and coke are heated to create pig iron

blow moulding a process where a thin tube of plastic (parison), gripped between two halves of a mould, is blown out to fill the mould using compressed air. Used for making bottles

brazing spelter an alloy of copper and zinc which melts a 875°C and is used as a filler

BSI British Standards Institute

CAD computer aided design

CAM computer aided manufacture

casting an object made by pouring liquid metal into a mould and allowing it to solidify. Acrylic and polyester resins can also be cast

CD-ROM Compact Disc – Read Only Memory

chuck a mechanical screw device for holding a tool in a machine, e.g. a drill bit

CIM computer integrated manufacture

clearance hole a hole made to enable to free passage of a screw-shank

CNC computer numerically controlled

coke a fuel source used in the production of steel

composites material made up of more than one base material, sometimes in layers or as a mixture

CNC computer numerically controlled machines are controlled using number values written into a program; each number is assigned a particular process

COSHH control of substances hazardous to health

converter a processing device used in the production of steel

countersink hole a hole made to receive the head of a screw and leave it level with the surface

cross links link together with chain molecules or other polymers with a transverse bond. The linking together of molecules on cooling to form a more rigid structure

cutting edge a sharpened edge used to cut

deforming process that allows material to change shape without changing its state, i.e. vacuum forming

die casting a process whereby metals are poured under gravity or injected under pressure into a metal mould

dimensional stability a material whose dimensions will not change when subjected to extreme environmental conditions such as moisture and humidity

drunk thread when the die of a screw is not aligned square to the axis of the rod

electrodes conductors through which electricity enters or leaves

electrolysis chemical decomposition by passing an electric current through a conducting fluid

EPOS Electronic point of sale

ergonomics the study of how products and environments are designed for human users

extrusion the process of forming uniform cross-sections of moulding used extensively for plastics and metals

fabrication the process of joining parts together

ferrous containing iron

fibrous made up of many thread-like fibres

fixture holding device similar to a jig that is fixed in place for immediate use

flotation a method of concentrating metallic ores using a soapy foam to trap the ore

flowchart a chart using symbols to show the sequence of a process

flux a substance used in welding or soldering to prevent oxides forming on the surfaces being joined

FMS flexible manufacturing system

former a base used to build up thin layers on a material to produce a desired shape or curve

fractionating column where crude oil is broken down into its various different products

Gantt chart a chart to show how a number of tasks processes can be completed in a given time, often simultaneously

Gauge a standard measure of the thickness of a screw, wire, sheet metal, etc.

hardness the ability to withstand abrasive wear and indentation

hardwood wood from a broadleaved tree

ingots cast rectangular blocks of metal

injection moulding a process where molten thermoplastic is injected under high pressure into a die cavity

invar an alloy of 63.8% iron, 36% nickel and 0.2% carbon. It has a low coefficient of expansion

jig movable holding device made to suit a single component in exact position

layout fluid a liquid used to coat metals prior to making cuts

laminating the process of joining sheet materials together to form solid sections or curved shapes

limestone hard rock made mostly of calcium carbonate; used in building materials

manufacturing cells a production system that incorporates a number of people and machines working together, being responsible for what is produced

mass production the production of a component or product in large numbers

mechanical properties properties of materials that are effected by an external force such as compressive or tensile forces

mechanical strength materials which are said to be strong, and resist or stand up to external forces such as compressive and tensile forces

meniscus the upper surface, or 'skin' or a liquid in a tube

MIG welding Metal Inert Gas welding – a relatively easy form of welding to carry out, commonly used in many school workshops

mock-up a model of a design in 3D, used for evaluation and testing

non-ferrous a metal that does not contain iron

one-off production a product required as a single item, such as a bridge or a football stadium

ore a natural material from which metals or minerals can be extracted

oxide a compound of oxygen and another element

oxidizes combines with oxygen to form a surface oxide

physical properties properties other than those which are effected by an external force such as density and conductivity

pilot hole a small hole used to guide a screw or thread

planed all round (PAR) timber that has been planed all round will have four smooth machined surfaces

plotter/plotter cutter a computer controlled output device for producing accurate lines or cuts on card or paper

presentation drawings drawings used to communicate a design in a suitable form for the client

prototype a model or product which has been made to be tested or trialled before being put into full production

quality assurance a policy or procedure written to ensure that a product reaches the customer to the correct specification

quality control systems put into place to check quality during manufacture – e.g. gauges, visual checks, etc.

realization taking a design on to the next stage after planning, that is, making a real product

reforming process involving a change of state within the materials used, i.e., from solid into liquid

rough-sawn timber which has come straight from the saw and has not been planed

scale of production the type of production – batch, mass, etc.

seasoning the process of reducing the moisture content in timber

shape memory alloy (SMA) a alloy that can be plastically deformed at a predetermined temperature, but that will return to its original shape after it has been heated

smelted metal extracted from its ore by heating and melting processes

softwood wood from a cone-bearing (conifer) tree

soldering joining metals with solder (a low melting alloy)

solenoid a cylindrical coil of wire which creates a magnetic field within itself when an electric current is passed through it. This means it can draw a core of iron or steel into itself

split die a tool used for cutting an external screw thread

sprue either a channel through which metal or plastic is poured into a mould or the metal or plastic which has solidified in a sprue; often holds pieces in model kits

stereolithography 3D modelling using lasers to solidify liquid plastics. Complex shapes can be produced – often called rapid prototyping

structural members the individual component parts that are subjected to forces, for example, chair legs being put in compression

swarf waste metal produced by turning, drilling, milling or thread-cutting operations

synthetic product made by chemical synthesis; usually to imitate a natural product

template a device used to mark out identical shapes

tensile strength ability to withstand tensile or stretchging forces

thermoplastic a type of plastic which softens under heat and can be re-softened many times

thermosetting plastic a type of plastic which once set cannot be re-softened or melted

thyristor a three-legged electrical switch that can be used as a latch, that is, it remains switched on even when the voltage is zero

tolerance the upper and lower limits of a dimension, e.g. 25mm +/- 0.5mm = 24.5/25.5mm

torsional forces forces which work to twist or untwist materials or structural members and components

turning the process of removing or wasting wood and metal on a lathe

user evaluation a method of evaluating a product by asking intended users what they think of the design and recording the results

vacuum forming a process using a thin plastic sheet which is formed around a mould using atmospheric pressure – used in blister packaging

veneers thin strips of hardwood (0.5mm thick) used for laminating onto cheap material to improve visual appearance

viscous sticky, slow flowing; used to describe a fluid which resists flow

wasting process that produces waste material by either cutting pieces out or cutting pieces off, e.g., chiselling, sawing

Index